THE IMPENDING WORLD ENERGY MESS

What It Is And What It Means To *YOU!*

ROBERT L. HIRSCH
ROGER H. BEZDEK
ROBERT M. WENDLING

We acknowledge the financial support of the Government of Canada through the Book Publishing Industry Development Program for our publishing activities.

Published by Apogee Prime, A division of Griffin Media

Printed and bound in Canada

The Impending World Energy Mess /Robert L. Hirsch, Roger H. Bezdek, Robert M Wendling
First Edition

The Impending World Energy Mess

By

Dr. Robert L. Hirsch

Dr. Roger H. Bezdek

Robert M. Wendling

With a Foreword by Dr James Schlesinger

Contents

Preface

Significant elements of the world energy enterprise are a mess. Energy is not a simple or easy subject in spite of what many would have you believe. Unfortunately, both politics and ideology too often cloud public discourse.

Delving into the physical and economic fundamentals illuminates the complications. This book is intended to help educate you on some of the most important energy issues; to show you why there is an extremely serious problem ahead in world oil supply; to explain why the promotion of some energy concepts border on deceit; and to suggest ways to get out of our energy mess.

Our intended audience is the intelligent, non-technical reader, as well as technologists, economists, and investors, who are interested in energy and the oil-related economic troubles that lie ahead. While our primary focus is oil, we delve into a number of other energy technologies and global warming. We make liberal use of graphs to display numbers, trends, and dependencies. In some cases, we have simplified technical complexities to minimize confusion, and we have done our best to explain our simplifications.

Our hope is that you, the reader, will end up more informed, so you can make more intelligent decisions in your personal lives and possibly even assume a more active role in energy-related discussions and politics.

We thank our families for their support as we have written this book. Our publisher, Rob Godwin has been a pleasure to work with, and their assistance is greatly appreciated. Finally, we thank the host of people who have contributed to our understanding of energy issues in recent years. Special thanks go to Dr. Jim Schlesinger for providing our Foreword. Others of note include the following: Kjell Aleklett, Steve Andrews, Peter Balash, Peter Blair, Colin Campbell, Ralph Carabetta, Doug Chapin, Frank Clemete, Debbie Cook, Jeremy Gilbert, David Gray, Mikael Höök, Sadad al-Husseini, Jim Katzer, Ken Kern, Rembrandt Koppelaar, Larry Kummer, Jean Laharrere, Jeremy Leggett, Fred Palmer, Mike Ramage, Matt Simmons, Chris Skrebowski, Dan Sperling, Randy Udall, Tom Whipple, and Jim Zucchetto. We apologize to those we have not listed. To all, we express our sincere gratitude for enriching our professional experiences.

While we have consulted various people on aspects of this book, it should be very clear that what is written here is our own work, except as noted, and is our sole responsibility.

While some of our messages are troubling and disheartening, we maintain our optimism that humankind will manage the world energy mess and end up stronger for the effort. But it will not be easy.

Dr. Robert L. Hirsch   Dr. Roger H. Bezdek   Robert M. Wendling

Foreword

This is an important book. Yet, despite the importance of the message, it will not be a welcome book – for its implications are unpalatable.

The ancient Greeks tell us that a Cassandra, despite the truth of her predictions, is doomed not to be believed. Over the years such has been the fate of those forecasting problems for our energy supply, for such forecasts provide discomfort - and political problems from which we prefer to avert our gaze.

Though we readily speak of oil as a finite resource, our tendency, having stated that as simple fact, has been to pass on to other subjects. Too frequently we are left with the thought that the problem of energy supply will magically disappear. For environmentalists the answer lies in conservation and renewables. More conservative observers think (or hope) that the free market will solve our problems. Indeed, even in the face of a supply shortage, markets will clear. However, the result in higher prices likely will imply reduced national output, the obsolescence of capital equipment, a decline of living standards, and severe political difficulties.

What are the hard facts that lie behind the "finiteness", in particular, of oil supply? One of the founders of the "peak oil" school, Colin Campbell, observed that in order to produce oil one must first discover it. That sounds like a truism or a bromide and it is too easily dismissed as a banality. Yet, when one considers the past and the prospects for future oil supplies, it provides a profound truth. Much of our oil production comes from oil fields that were discovered in the 60s and 70s of the last century. In recent decades we have produced far more oil than we have discovered. In fact, for every three barrels of oil that we have produced, we have found roughly one barrel of new oil.

The challenge may be simply, if dramatically, put. The decline rate from presently producing fields is roughly four million barrels per day. Given the need to replace that production and to increase producing capacity to cover the growth in expected demand, would require the discovery and development in the next quarter century of the equivalent of five Saudi Arabias. No one really expects to find five Saudi Arabias. To this date, we have only found one.

Concerned analysts, including the authors of this book, believe that world oil production will remain on the fluctuating plateau it has been on since 2004 for another 2-5 years before it begins to decline. Optimists believe that we have several decades more before the final oil production plateau. The industry generally does not subscribe to the peak oil theory. For some in the industry the problem is political, the so-called "above ground" problem, rather than geological. The industry generally contends that there are ample resources, but the nations in which the resources are to be found will not allow efficient exploitation. Nonetheless, as one former CEO has told me, "of course I know there will be a peak, I just don't know when it will come".

The authors provide a detailed analysis of the many proposed alternatives that might provide physical mitigation for shortages of liquid fuels. It should be noted that the energy problem is basically a problem of liquid fuels. Starting in the early 1970s, the energy problem has all too frequently been treated as if it could be solved by increasing electric power production. Indeed power production could, in principle, be substantially increased, but it would not solve the problem of providing the liquid fuels on which transportation worldwide is now dependent.

Some readers will be surprised, even distressed, by the skeptical treatment of global warming. Nonetheless, between the embarrassments suffered by the leakage of e-mails from the Climate Research Unit at East Anglia University, and that the climate models have consistently provided inaccurate forecasts of temperature changes, one should - at a minimum - read the skepticism expressed with an open mind. Perhaps even more importantly, one should find it puzzling that it remains the prevailing fashion to devote so much attention to what remains a hypothesis regarding the impact of the release of greenhouse gases on temperature at century's end about which we can do little, given the fuel choices in the developing world, notably China.

Why is it puzzling? First, the called-for reduction of greenhouse gas emissions seems impossible to achieve. The Energy Information Administration, for example, now projects a 50% increase in carbon dioxide emissions over the next quarter century. More importantly, the projected rise in temperature for the year 2100, 90 years away, remains rather distant and quite theoretical.

At the same time, astonishing little attention is paid to the inability to increase oil supply, as demand rises within the next decade or two. Perhaps the reason we devote attention to the former is that the latter is a near term certainty which we prefer not to acknowledge.

Readers of this book may question individual calculations by the authors, but they need to absorb the overall message of the analysis regarding the inability in the decades immediately ahead to increase production of liquid fuels, as demand rises.

James Schlesinger

Dr. James R. Schlesinger was the first U.S. Secretary of Energy. He also served as Chairman of the Atomic Energy Commission, Director of Central Intelligence, and Secretary of Defense. He is currently Chairman of the Board of MITRE Corp., a consultant to the Departments of Defense and State, and a member of the Defense Policy Board and the International Security Advisory Board. Among other honors he is a fellow of the National Academy of Public Administration and a member of the American Academy of Diplomacy.

I. Introduction

The world is headed for serious trouble in energy. Why? How will it impact you? How can you minimize damage to your personal life?

Our purpose is to give you information you can use as the manager of your own economic future. Some of what is involved is not easy reading and may cause distress. If the subject was simple, the U.S. and the world probably would not be in its current predicament.

The three authors of this book have been involved in various aspects of energy for a total of over 100 years, and we believe that we understand many of the realities. We have learned a great deal over our careers. Our aim is to provide facts, where they are clear, and to provide seasoned judgments, where the meaning of information requires interpretation.

In the course of our involvement in energy, we have encountered the following roadblocks:

- **Ignorance of energy basics,** which is unfortunately widespread. Energy has been available reliably and relatively inexpensively for most in the developed world when needed, so people have spent time and effort on other things.

- **Incompetence**, for all the reasons it exists.

- **Intellectual rigidity**, where people are so tied to history and their training that they fail to recognize that different technologies require different thinking.

- **Short-term thinking** that limits attention to the immediate and minimizes serious consideration of longer-term problems.

- **Self-interest,** often associated with a person's employment. If a company or environmental organization might suffer if certain realities were publically understood, then self-interest often dictates less-than-full disclosure or smoke screen lobbying through the media or with policy-makers.

- **Conspiracy** among people and organizations protecting their common turf. In this case, self-interest often leads to banding together to obscure certain inconvenient truths.

Finding fault may be satisfying to some people but not to us; our interest is in seeing bad decisions and mistakes corrected as fast as possible, so that energy-related economic damage can be minimized.

II. What This Is About

Energy is essential to our everyday lives. Human muscle power is capable of producing roughly 35 watts of energy on a continuous basis, which is a small fraction of one horsepower. Think of that when you flip on a 100-watt light bulb or drive your 100-300 horsepower car.

The energy we use in our everyday lives takes a number of forms. Two of the most important are liquid fuels and electricity. For many of us, gasoline is the liquid fuel that we use the most. Almost all cars, trucks, trains, buses, ships, and airplanes are powered by liquid fuels, primarily derived from oil. Those vehicles deliver our food and consumer goods; they enable local, national and international commerce; and they move us to and from work and recreation.

Electricity is all around us in our homes and almost everywhere we go. It is in the form of lights, computers, television sets, refrigerators, cell phones, air conditioning, washers and dryers, etc. Wires bring electric power to us in most places where we live, work, and recreate. Few of us can conceive of living without it!

In thinking about energy, it's important to recognize that these two energy forms are very different and rarely interchangeable. It would be unrealistic to think of powering computers, television sets, or light bulbs with oil; they are built to operate on electricity. By the same token, almost all existing cars, trucks, and airplanes are built to operate on liquid fuels, not electricity.

In this book the focus is on oil and on electricity. Among the most important points are the following:

1. Oil is a finite resource that we have been producing at ever increasing rates to fuel our expanding economies. World oil production will soon reach a maximum rate of production, if it hasn't by the time that you read this book. Thereafter, there will be less and less oil available each year. The result will be growing shortages, rapidly escalating oil prices, and economic distress.

2. Oil shortages impact economies. What happened during the brief oil shortages that occurred in 1973 and 1979 gives us insights into what lies ahead. Understanding sequences of events and impacts will help you to understand the likely magnitude of the ensuing problems, and it will help you to better prepare yourself for what's coming.

3. Differentiating between oil resources and reserves is essential; they

are very different. Big new oil discoveries are important, but they will not have the large impact the numbers might imply.

4. The most significant options for mitigating the upcoming world oil shortage are described. Because the use of oil is so pervasive and the volume of consumption is extremely large, there will be no quick fixes or "silver bullets" like the ones used to hasten our emergence from the recession of 2008-10. Indeed, a worldwide crash program of oil shortage mitigation will require a long time to take hold, because the decline of world oil production will have a head start.

5. Renewable energy sources are not the ready solutions to our energy problems, as many profess. Each renewable technology has shortcomings, just like every other energy option. In some cases, the shortcomings are a matter of high total costs, while in others, the problems are ones of simple inadequacy. In choosing practical electric energy technologies, more informed choices must be made, or our energy systems will fail us before long. Unfortunately, idealism and wishful thinking have often clouded practicality, especially in the cases of solar and wind energy.

6. Governments can make costly, time-wasting mistakes in picking energy winners and losers. Politicians have often ignored the realities that the marketplace must respond to. The history of corn-ethanol shows how poorly informed government actions can be wasteful and costly.

We consider ourselves optimistic pragmatists, not "Armageddonists." There can be a bright future ahead after we overcome our impending energy problems, but it will not be easy. We believe that the economies that emerge after our immediate energy problems are solved will be more efficient and resilient.

III. Thinking About Oil

A. Introduction

Look around you. Oil and its uses are everywhere. Outdoors, there are cars running on gasoline derived from oil. Most buses and trucks run on diesel fuel derived from oil. Airplanes run on jet fuel derived from oil. Roads are often made of asphalt from oil. Oil-derived liquid fuels power most boats, ships, trains, lawnmowers, jet skis, motorcycles and a myriad of other machines.

In addition to fuels, oil provides a large number of everyday products, including detergents, plastics, lubricants, solvents, roofing materials, antifreeze, bandages, CD's, antiseptics, insecticides, pharmaceuticals, perfumes, refrigerants, tires, aspirin, shampoos, enamel, paints, dyes, insect repellant, trash bags, polyester, nylon, contact lenses, shaving creams, toothpaste, paints, soaps, epoxy, etc.

Consider the construction of buildings, bridges, and homes for example. Building materials have to be transported to construction sites by gasoline or diesel-fueled trucks and trains, often over great distances. At construction sites, gasoline and diesel-fueled machines move construction materials around and into position, and the people doing the work get to work in vehicles fueled by oil-based products.

Gasoline moves many of us to and from work every day. Oil-derived fuels deliver most of our packages around town and around the world. Fuels from oil bring us goods from overseas, and they move our exports to others. Indeed, if you spend some time thinking about how economies use oil-based products and take some time to identify oil-dependent activities around you, it is easy to understand why oil products are so essential to our very existence.

In the past, oil has been generally plentiful and low-cost. In the not too distant future, world oil production will change in a way that has never happened before. It will irreversibly begin to decline, causing growing shortages and oil prices to increase dramatically, forcing most of us to face serious, fundamental problems that will affect our very existence.

Beyond our discussion of oil, we delve into electric power technologies and describe the strengths and weaknesses of some of the most widely discussed electric power generation

> *"It is easy to understand why oil products are so essential to our very existence."*

> **"All energy options have strengths and weaknesses. Practical choices have to be made, and compromises are always required. Recently, idealism and ignorance have often interfered with pragmatism."**

technologies. All energy options have strengths and weaknesses. Practical choices have to be made, and compromises are always required. Recently, idealism and ignorance have often interfered with pragmatism. This will have to change.

Our hope is that this book will help you to better understand some of the important realities in energy and give you some of the information that you need to adjust your own life so as to minimize your risks and possibly even benefit from what lies ahead.

B. Oil is energy but not all energy is oil.

It is essential to recognize that different machinery and equipment are built to operate on specific forms of energy. Computers run on electricity; obviously, you cannot operate them on gasoline. All but a few automobiles run on gasoline or diesel fuel; they cannot be plugged into an electric outlet and powered by electric power, because they are not built to operate that way.

Electricity powers our appliances, computers, cell phones, subway trains, and a myriad of other devices. We plug them in to operate them; we do not pour liquid fuel into them.

Almost all liquid fuels are hydrocarbon-based, because of their high energy content, their convenience, their relative ease of handling, their safety, and their relatively low cost. Almost all liquid fuels in everyday use are refined products originating from oil. The highest volumes of liquid fuels derived from oil are gasoline, diesel fuel, jet fuel, heating oil, and bunker fuel for ships.

Large civilian and military airplanes are powered by jet fuel. Practical use of alternate fuels in airplanes is a distant dream, requiring wholly new designs, construction, and fueling infrastructure. And the list goes on and on.

Switching the energy we put into almost all machinery and equipment

> **"Switching the energy we put into almost all machinery and equipment is impossible, because of the way they were built."**

is impossible, because of the way they were built. To run automobiles on electricity, for example, requires production of a new type of automobile, specifically engineered and constructed to be electrically powered. To operate heavy trucks on natural gas, as some advocate, we must either undertake an extensive retrofit of existing vehicles or we must design and build new ones from scratch.

Many long distance trains operate on diesel fuel, and many short distance trains operate on electricity. To switch all trains to electric power, we would need to build new locomotives, add electric lines and third rails to train tracks, and construct new electric power plants to supply the needed power. These and other transformations can be done, but they will take a long time and vast sums of money before significant national or world-scale change is achieved. The effectiveness of such changes is dependent on the extent and speed of implementation. These observations may seem obvious, but they are often ignored by people who claim that large changes in our energy system can be made quickly.

Many energy-form changes are attractive for the long term, but most equipment owners are unable to discard what they have until the end of their equipment's useful life. To think it easy or inexpensive to quickly switch energy form for a given piece of equipment is naïve. Overnight machinery change is a fantasy. We cannot afford to simply discard capital stock that has years of useful life remaining. Change typically does not occur until near the end of the equipment's useful lives.

> **"Overnight machinery change is a fantasy. We cannot afford to simply discard capital stock that has years of useful life remaining. Change typically does not occur until near the end of the equipment's useful lives."**

Because of this simple fact-of-life, the world's capital stock of oil product-fueled equipment will be with us for close to their useful lifetimes. Thus, when dealing with oil supply problems, talk of windmills saving us is

> *"Oil is energy, but not all energy is oil. Almost all equipment built to operate on oil products cannot operate on electric power."*

without merit. Windmills produce electricity, not liquid fuels. Oil is energy, but not all energy is oil. Almost all equipment built to operate on oil products cannot operate on electric power. Those who talk of quickly solving our oil problems with wind or solar cells are showing ignorance of the fundamentals or worse.

In the longer term – decades – various types of equipment can be redesigned and built to operate on electric power. For that to happen, there will have to be a demand for such alternate equipment, which means it will have to perform as well or better than that which currently exists, and it will have to be economic to buy and operate. We have no doubt that there will be significant changeovers of equipment power from liquid fuels to electric power in many applications in the future. Electricity's share of world energy consumption has been gradually increasing for decades, and this trend is forecast to continue. However, as we show later in this book, it will require decades before there will be a meaningful impact on liquid fuel demand on a national or international scale.

C. Basics of the Oil Business

Oil, often called petroleum or crude oil, is an incredibly complex mixture of hydrocarbon molecules – molecules composed of hydrogen and carbon. Oil was created tens of millions of years ago by the geological burial of plant life, usually in lakes, rivers and oceans. Over time, these materials were buried ever deeper; they were cooked by the earth's heat and compressed by various natural forces. The end-result was an array of light and heavy oils and natural gas. The variations in physical properties depend on a wide range of complicated phenomena. Their natural forms and their impurity content do not make them useful in most modern machinery, so considerable processing is required, which is accomplished in refineries.

Consider the exploration for and production of oil and natural gas. Both fossil fuels are found at varying depths in parts of the world where the right conditions existed in the distant past; oil and natural gas are not found everywhere. Both oil and gas are found in reservoirs -- containers that held them over millions of years, inhibiting them from leaking up to the surface of the earth. Oil and natural gas are coupled in this discussion because they are often found together. Many oil reservoirs have "gas caps" that sit atop the oil.

> ## "Oil and natural gas are not found everywhere."

Oil and natural gas reservoirs are not underground cavities; they are porous rocks with oil and gas in the pore spaces. These "reservoir rocks" are covered by impermeable cap rocks that inhibit oil and gas from migrating out.

Finding oil and natural gas is very difficult, particularly since the world has moved beyond the easier-to-find resources to the much more difficult. A significant effort goes into deciding whether a new area for exploration is worthy of serious investment. However, nothing is known for certain until a well is drilled to determine if hydrocarbons are present. Wells that fail are called "dry holes." Sometimes a well will find hydrocarbons but not in sufficient quantities to justify further development. Useless wells are plugged with cement and abandoned.

When economically viable oil and natural gas are found, many more wells are drilled to provide multiple paths to extract the hydrocarbons. Those wells must be connected by piping that leads to facilities that do some initial processing, yielding oil and/or natural gas that must then be sent by other pipelines to either a refinery, if it is close by, or to tanker ships, which can transport the materials to more distant refineries.

Raw crude oil must be processed to produce the myriad of finished products used in our everyday lives. A modern refinery covers a large area with a vast array of interconnected processing facilities that separate oil molecules into rough product streams, called "fractions," that are then further treated to produce those marvelous fuels, chemicals, plastics, pharmaceuticals, and other products that enrich our everyday lives and power our economies. Refinery processes perform functions such as molecular separations, composition changes, and impurity removal, so that the final products fit the required specifications that provide optimum performance and environmental acceptability.

After refining is complete, each product is transported to markets (gasoline stations or other distribution facilities) or to yet other refineries for further processing into other products. A huge system has grown over time to meet the ever-increasing demands of modern industrial economies. As demands for higher product quality have increased, the refining industry has risen to the challenges, sometimes kicking and screaming. It is not that refiners do not want to provide cleaner products; they just do not want to spend money on new equipment, because refining profit margins are often very small, much lower than in other industrial activities.

As technologies advanced, all of the oil exploration, production, transportation, refining, and distribution activities have become better optimized, increasingly safer, and increasingly more environmentally friendly. Nevertheless, like any large-scale industrial activity, there are periodic accidents, spills, explosions, etc. In many cases these events were tragic and heart rending when people or wildlife suffer. It is essential to remember that nothing is perfect – not any of us, or the things we do. However, considering the huge volumes of hydrocarbons that are processed every minute of every day, the record is remarkably positive. Every energy technology has its dangers and shortcomings, as we describe later in this book. Often new technologies seem cleaner or better than what we have, but over time, we see that all have weaknesses.

> ## *"Every energy technology has its dangers and shortcomings."*

D. Oil Numbers in Perspective

Most technical people do not comprehend the size of world oil consumption and the fact that it takes a long time to build major new facilities. One misconception relates to how fast things can be changed. Many people are conditioned by what has happened in computers and electronics, where remarkable changes in capabilities and costs have been implemented over the span of months and years. Those rapid changes are largely a function of the fact that electronics and microcircuits are inherently very small. "Small" allows for changes to be made and implemented quickly.

Energy facilities are inherently very large-scale. Mega's (millions) and giga's (billions) abound. For instance, the world consumes roughly 85 million barrels of oil every day. On an annual basis that's over 30 gigabarrels – 30,000,000,000 barrels. If we were to lay oil barrels end-to-end, a day's world oil consumption in barrels would circle the world twice. A year's worth of oil barrels would circle the globe about 700 times. Suffice to say, BIG!

Recently, there have been a few new giant oil field discoveries with upwards of ten billion barrels of oil (10 gigabarrels), which sounds enormous. However, once you recognize that many decades are required to extract that oil, it becomes easier to understand that at maximum production, a 10 billion barrel oil field may produce roughly a million barrels per day during its best years, after which it will produce less and less in each ensuing year, often for decades. Thus, during its best years, a 10 billion barrel oil field would contribute just over 1% of recent world oil demand.

> ## *"Energy facilities are inherently very large-scale."*

Recognize also that huge oil fields take a long time to bring into operation, because they represent massive, expensive industrial undertakings. Consider, the Tupi giant oil field that is offshore Brazil, discovered a few years ago. Estimates are that it might contain between 5-8 billion barrels of oil equivalent, which is industry jargon for the oil energy of both the oil and natural gas that would be produced together. To fully develop the Tupi field to full production may cost of the order of $50 billion dollars. If all goes well, Tupi will produce about 0.22 million barrels per day plus sizable amounts of natural gas in 2013, increasing to about 0.6 million barrels per day oil in 2015[1]. Development will not be quick because it is very difficult to produce in deep water, and Tupi's complex geology presents significant technological challenges. Large scale means expensive and slow.

Some other numbers: 1% of world oil production is about 850,000 barrels per day. If we were to produce that much oil from Coal-To-Liquids plants, it would take more than eight 100,000 barrels per day facilities, each costing over $5 billion apiece, approaching half a trillion dollars for a 1% contribution!

No matter how you cut it, the scale size that we are dealing with in world oil production is enormous.

As we will touch on later, proponents of processes to turn plant material (biomass) into liquid fuels do not typically use barrels per day when they describe their favored technologies; they talk about millions of gallons per year, which sounds large. However, the standard for comparison in the liquid fuels business is millions of barrels per day. A barrel contains 42 gallons, so a million gallons per year of biomass liquids is about 65 barrels per day, which is not very impressive. Ethanol from corn has a lower energy content than gasoline, which it is supposed to substitute for, so a million gallons of ethanol per year has the energy content of just over 40 barrels per day of gasoline, which is an infinitesimally small portion on the world scale.

> ## *"Large scale means expensive and slow."*

IV. Understanding Oil Fields

Oil fields consist of one or more oil reservoirs. An oil reservoir is a deeply buried rock formation composed of porous rock, which holds oil in its pores, and a cap rock on top that acts as a container. Think of a bowl of water; the water wants to drain downward but is contained by the bowl. In an oil reservoir, the oil wants to escape upward, due to various physical forces, but the cap rock acts like an inverted bowl to hold the oil in place. Like a bowl of water, an oil reservoir contains a fixed amount of oil, which we drain, when we drill oil wells into it, releasing oil to flow to the earth's surface.

In a bowl, water sits in place, held by gravity. In an oil reservoir, oil is trapped in the rock pores, like water in a sponge. However, unlike a sponge, reservoir rock cannot be squeezed to release the oil. Rather, oil must flow between adjacent pores to find its way to oil wells, where it can flow out.

Oil reservoirs were created by a remarkable confluence of physical and geological events over the course of millions of years. Oil reservoirs are hard to find because they are typically located a mile or more below the surface of the earth, and we have few ways of identifying their presence. Thus, oil finding, called *exploration*, is a difficult task, which is getting progressively more difficult, as the oil industry moves beyond the easiest-to-find oil to the more difficult-to-find.

A number of oil reservoirs are often found close together. Such a grouping is called an *oil field*. Adjacent oil reservoirs in an oil field are typically developed together for convenience, efficiency, and economics, because field development is a complex, expensive endeavor.

In the simplest terms, the finding and development of an oil field involves the following steps:

1) **Exploration** to locate oil fields. Oil fields are not found everywhere in the world, and the largest accumulations are found in relatively few countries. An exploratory well drilled into a potentially productive formation provides the first real indication of the presence of oil. Unproductive wells are called *dry holes*.

2) **Appraisal** follows a successful exploratory well and involves

"In an oil reservoir, oil is trapped in the rock pores, like water in a sponge."

drilling additional wells to determine if the oil that is present is economical to produce. If little or no additional oil is found, the area is abandoned in spite of having expended a large amount of time and money.

3) **Development** is the process where more wells are drilled and connected by pipes that carry oil to processing facilities, which are then connected to refineries, sometimes through long pipelines or by shipping. Development often continues for decades. New wells are continually drilled and connected; old wells must be kept in good operating condition and sometimes re-drilled; facilities must be maintained and upgraded; wastes must be properly managed and disposed of, etc.

At some point the natural pressure in oil reservoirs becomes too weak to keep oil flowing at high rates, so alternate means of pressuring reservoirs are used. These often include injecting large amounts of water to push or float oil to producing wells, or enhanced oil recovery, which involves other techniques for pressurizing and moving oil.

4) **Plug and Abandon** marks the end of an oil field. It involves
   1) sealing the wells, usually by injecting cement to insure their isolation;
   2) removing surface facilities; and
   3) environmental cleanup.

A production profile of a typical giant oil field is shown in Figure IV-1, where the time horizon extends over many decades, sometimes approaching a century. Figure IV-2 shows production profiles of four oil fields that are past their maximum production levels. As the figures illustrate, production profiles can vary considerably. For giant fields, a production plateau is usual, while for smaller fields, the production profile can be relatively sharp.

In all cases, a point is reached in the lifetime of every oil field when production goes into decline. This is a fundamental fact of nature and occurs because the recoverable oil in every oil field is finite, just like the liquid in a bowl or pail or tank. The long production decline is the result of a variety of physical factors, which inhibit oil flow through microscopic pores in reservoir rocks.

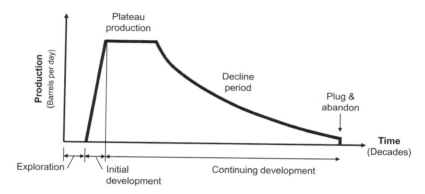

Figure IV-1. Oil production profile of a typical, large oil field. After years of exploration and the drilling of successful appraisal wells, large-scale development begins. Oil production increases to a maximum and may stay flat on a production plateau for a number of years, sometimes a decade or more. Then, despite all human efforts, production decline sets in and can continue for many more decades.

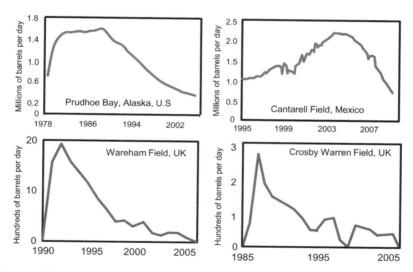

Figure IV-2. Oil production profiles for four oil fields of varying production outputs. Plateaus are seen in larger fields, while smaller fields often show a sharp peak. Note the production levels: The giants are in millions of barrels per day while the dwarfs are in hundreds.

V. Oil Reserves & Production – Things You Need to Understand

A. Oil Reserves

As already noted, an oil field is an area with one or more oil reservoirs, sometimes close together, sometimes somewhat separated. The total amount of oil in an oil field is called its **resource** – all that is there. Because of the complexity of reservoir rock structures, significant portions of the oil resource are trapped in closed mini-structures or otherwise not practically accessible. The result is that the amount of oil that can be practically produced is a fraction of the resource, which is called its **reserves**, typically 25-40% of the resource.

Why not do what is necessary to recover all of the oil resource? The answer is that it is practically impossible, because we would have to drill wells into all kinds of small structures that are impossible to identify from the earth's surface, often miles away.

To help you understand why getting most of the oil resource is essentially impossible, you can gain some insight into geological complexity in the following way. Next time you drive through high hills or mountains where roads are cut through, look at the exposed rock structures.[2] Where the road cuts are deep, you will often see the complexity of the rocks with layers going up and down at different angles, layers of rock pinched off, rock displacements next to fractures, fractures that are cemented closed, etc. What you see in these situations gives you some clue as to what exists in oil field rocks.

Another insight comes from considering oil company motivations. Technologists in oil companies are highly educated and typically have decades of experience working on real oil fields. They know what can be done, how to do it, and what the costs would be. Company managements are in business to make as much money as possible from the oil fields that they develop. Profit can be increased, if they can produce more oil from their oil field investments, which are often huge. If there were practical ways to extract more of the resource with modest to significant incremental investments, they would do it. The fact that they often cannot extract more of the resource tells you that it is extremely difficult, very costly, and often practically impossible.

As noted, the quantity of oil that can be produced over the life of an oil field is known as its **reserves**.[3] Estimating reserves is very difficult, because of reservoir rock complexity, which is better understood as more wells are drilled to develop an oil field. Accordingly, revised estimates of reserves are periodically made.

While reserves are important, we live on oil that flows to where we can use it. The value of reserves in the ground is that they can be produced. Until they are produced, they have no practical value to our everyday lives. People sometimes think that reserves mean production, which is not necessarily the case. For instance, Venezuela has large oil reserves but falling production due to mismanagement. Iraq is thought to have very large oil reserves, but war and its aftermath have so far limited production.

Besides geological complexity, estimates of oil reserves are dependent on the expected price of oil over the lifetime of the field. Why? Because when high oil prices are expected, an operator can justify drilling more wells and spending more on maintenance and enhanced recovery. If low oil prices are forecast, less investment is justified, resulting in less ultimate oil production, meaning lower reserves. Reserves estimates are thus oil price dependent to a degree, but the range of variations are rarely greater than 10%, often less.

In some ways, oil reserves are like inventory, defined in the dictionary as *"The quantity of goods and materials in stock."* In the case of boxes in a storeroom or liquids in a tank, inventory can be removed at a rate dictated and managed by humans, e.g., ten boxes per day, so many gallons per hour, etc. (Figure V-1). However, the inventory analogy is limited and can lead the non-expert to incorrect conclusions about oil field behavior.

In the early stages of oil production from a new oil field, an operator has a good deal of control over oil production rates up to a maximum, determined by geology, investments, surface conditions, and other factors. When oil production begins to decline, the oil withdrawal rate is dictated primarily by nature -- the physics and chemistry of decreasing oil flow through porous reservoir rocks. This fact has fundamental ramifications for oil production management in a field, a country, and the world, as will become obvious in subsequent discussions.

"While reserves are important, we live on the oil that flows to where we can use it."

B. Are oil reserves estimates always believable?

The foregoing description of oil reserves estimation is the way things would work in an honest world. The problem is that in a number of cases, honesty is not the operative criterion. Sometimes, oil field operators or politicians demand reserve estimates that are much higher than the facts justify. In those situations, engineers face a choice between giving their bosses what they want or losing their jobs. Unfortunately, there are a number of important situations where such pressures exist, so publically available reserves estimates are not always reliable; in some cases estimates are outright exaggerations, always on the optimistic side.

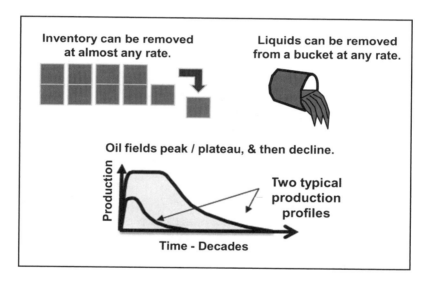

Figure V-1. Oil reserves are not like normal inventory. Boxes in storage or liquids in a bucket can be withdrawn at a rate determined by people. After the onset of production decline in oil fields, the withdrawal rate is dictated primarily by nature.

As an example, consider the situation in OPEC in the mid 1980s. OPEC decided to change its oil production quota formulas to take into account the reserves that its members claimed. Higher reserves meant higher production quotas. Does this sound like a situation ripe for gaming? As described in Wikipedia:

"There are doubts about the reliability of official OPEC reserves estimates, which are not provided with any form of audit or verification that meet external reporting standards.

"Since a system of country production quotas was introduced in the 1980s, partly based on reserves levels, there have been dramatic increases in reported reserves among OPEC producers. In 1983, Kuwait increased its proven reserves from 67 Gbbl to 92 Gbbl.[4] In 1985-86, the UAE almost tripled its reserves from 33 Gbbl to 97 Gbbl. Saudi Arabia raised its reported reserve number in 1988 by 50%. In 2001-02, Iran raised its proven reserves by some 30% to 130 Gbbl, which advanced it to second place in reserves and ahead of Iraq. Iran denied accusations of a political motive behind the readjustment, attributing the increase instead to a combination of new discoveries and improved recovery. No details were offered of how any of the upgrades were arrived at.

"This reality makes for real difficulties in calculating how much more oil is in the ground to be recovered, which makes the estimation of the date when world oil production goes into decline difficult to estimate."

Those dramatic increases in reserve estimates occurred during a period when relatively few new oil field discoveries were made in the OPEC countries, so the revisions could not be based on new finds. The result is that the world was and still is faced with uncertainties regarding the amount of actual oil reserves in OPEC, which is believed to have the largest reserves in the world.

Furthermore, within most OPEC countries, oil reserves estimates are considered state secrets, and few countries have allowed outsiders to audit their published reserves claims. Some country estimates are obviously suspect. For example, since the magic jump in Saudi Arabian oil reserves in 1988, Saudi Arabia has stated that its reserves have fluctuated very close to 260 billion barrels every year, in spite of the fact that Saudi production was roughly 3 billion barrels per year. It is plausible that some reserves re-estimation and some new discoveries might balance production over a year or two or maybe even three. However, Saudi Arabia has not had any major new oil field discoveries since the early 1980s. Thus, to claim an essentially perfect balance for over 20 years is asking us to believe something that is highly improbable.

OPEC has long had an interest in the world thinking that there will be no world oil production limits in the future. Why? If huge supplies were accepted as true, importing countries would not be motivated to seek alternatives to imported oil, and OPEC could continue to sell its oil at stable or increasing prices. Self-interest is not hard to understand, so OPEC's motivation should not be a surprise.

> ## "Self-interest is not hard to understand, so OPEC's motivation should not be a surprise."

Some, Matt Simmons[5] for example, have challenged Saudi oil claims. Others, who are conceivably in a position to do so, have not. Why? Consider some of the possible challengers and their circumstances:

1. Many oil service companies provide oil field services to OPEC countries. As such, they may have insider information about oil realities in regions where they operate. They could raise questions about OPEC member claims, but that would involve biting the hand that feeds them.

2. Some of the major oil companies formerly (or currently) operate in OPEC countries. Those companies and others hope for OPEC country business in the future, so it is not in their self-interest to challenge OPEC claims and get frozen out of future opportunities.

3. The U.S. government (USG) has had a special relationship with the Saudis for decades. In past tight oil supply situations, it is rumored that the Saudis accommodated U.S. special requests by periodically providing relief. The USG recognizes that there might be more such situations in the future, so why would the USG risk antagonizing the Saudis by challenging their reserves claims? U.S. security organizations may have special insights into what the Saudis or OPEC actually has, but they are not made public.

4. Oil field consultants could raise issues and some have. However, many consultants receive financial support from OPEC countries or might in the future, so they have little interest in antagonizing existing or potential clients.

5. A number of oil industry publications could raise questions, but their business is selling their journals and magazines, so why risk losing lucrative markets?

6. The International Energy Agency (IEA) could challenge OPEC, but if it did, they could lose sources of information, which could be to their disadvantage. About the most that the IEA has said is that it has no reason to doubt what OPEC has claimed.

So the world is moving towards greater dependence on OPEC oil production with little substantive knowledge of what they really have. OPEC is in effect asking oil importers to "Trust Us." This is not acceptable and represents a huge risk for oil importers.

But the situation gets worse. Not only are we ignorant of OPEC's actual reserves, we are dependent on OPEC making the investments needed to bring whatever oil they have to market. Huge investments are required to expand production on the scale needed to satisfy growing world needs. Will those investments be forthcoming? Is it in OPEC's best interest to spend ever more money to increase oil production, which would hold down world oil prices, which will mean less income? Obviously, it is not.

> *"OPEC is in effect asking oil importers to "Trust Us." This is not adequate and certainly represents a huge risk for oil importers."*

Counterbalancing the uncertain publically available data is the work of a number of knowledgeable oil experts. Thus, there are world oil reserves estimates that are reasonably fact-based and worthy of serious attention. These estimates paint a gray or dark picture, as we will discuss later.

What about the reserves estimates of the major International Oil Companies (IOCs) such as ExxonMobil, Chevron, ConocoPhillips, Royal Dutch Shell, etc.? All oil companies that operate in the U.S. or are listed on U.S. stock exchanges are required by the U.S. Securities and Exchange Commission to conservatively estimate their oil reserves. The motivation is to minimize the risk of companies claiming more than they have, thereby stimulating inappropriate stock valuations. Thus, reserves estimates from oil companies operating in the U.S. have stood up well over the decades, but every so often there are exceptions. For instance, a few years ago Royal Dutch Shell had an internal problem that resulted in skewed reserves estimates. As a result, Shell had to substantially reduce their reserves estimates and pay heavy fines for their poor management.

C. Regional Oil Production

Oil output in producing countries comes from oil fields in varying stages of their lives – some starting up, others on plateau, and still others in decline. A simple illustration of how production patterns can add up is shown in Figure V-2, where hypothetical production from a dwarf, a giant, and a super giant field are shown. The purpose is to illustrate how production can add up and change over time.

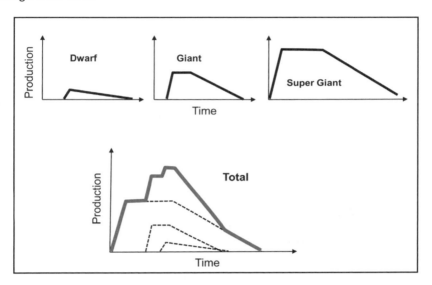

Figure V-2. Illustration of how the production from three different sized oil fields adds up to a total. The timing and sizes are arbitrary.

Consider the history of U.S. oil production, shown in Figure V-3. First production began in 1859 and remained relatively small until the early 1900s, when demand due to automobiles began to rise. Production from the U.S. Lower 48 states rose for decades and then reached a sharp maximum in 1970, after which it went into a long decline. Alaskan oil production came on stream in the late 1970s and rose quickly, offsetting some of the oil production decline in the Lower 48. Nevertheless, declines are a fact of nature, and new U.S. oil discoveries could not make up for the declines inherent to Lower 48 production.

Similar patterns of national oil production ramp-ups and declines have been repeated around the world. In 2005 the Royal Swedish Academy – they bestow the Nobel Prizes in chemistry and physics – did a study of world oil production and noted with alarm that 55 of the world's 65 largest oil producing countries had reached maximum oil production or were in decline. Since then, Mexican oil production has gone into sharp decline, and some analysts think Russia may be about to join the declining production club.

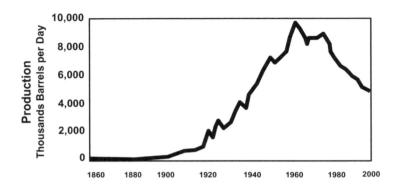

Figure V-3. US oil production from 1859 to 2000. Production rose dramatically to a sharp peak in 1970 and then went into decline until Alaskan production provided an important boost. Thereafter total production again went into a long decline that continues today.

With so many countries experiencing oil production decline and the world demanding ever more oil for future growth, the world is in a situation where fewer and fewer oil producers are being asked to provide ever more production. At some point that situation will break down.

D. Will higher oil prices and technology save us?

There is a hard core of **oil problem skeptics**, who tell us not to worry, because higher oil prices and technology will save the day. These skeptics do not seem to understand depleting resources, or they think that the resource is much larger than do most geologists. The history of U.S. oil production provides some hard facts that discredit the high oil prices/technology-will-save-us school of thought.

Recall that in 1973 OPEC embargoed oil exports to the U.S., which suffered its first sudden oil shortage. World oil prices increased dramatically, as a result. In 1979 the Iranian revolution led to a second major oil shortage, again causing oil prices to skyrocket. Oil prices stayed high until 1986, when they declined precipitously because of a world oil supply glut. So

> *"The world is in a situation where fewer and fewer oil producers are being asked to provide ever more production."*

between 1973 and 1986, the U.S. saw much higher oil prices. If increased oil production was there to be had, U.S. production should have increased dramatically, according to the high oil price thesis, but it did not.

On the technology side, the technologies used in oil exploration and production began a dramatic upswing in capability and effectiveness in the early-mid 1970s. Technological improvements have continued ever since. If the technology-will-save-us crowd was correct, then the later 1970s and early 1980s should have seen a technology response in U.S. oil production, but it did not.

What actually happened in U.S. Lower 48 states in the 1970s and beyond is shown in Figure V-4. Oil production continued the decline that started in 1970 in spite of higher oil prices and much improved oil field technologies. The combination did not reverse the inevitable decline. Mother nature did her "thing" in spite of some human postulates to the contrary.

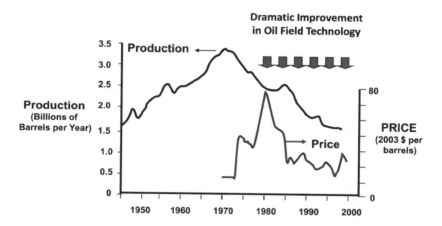

Figure V-4. Oil production in the U.S. Lower 48 states. Much higher oil prices and greatly improved oil field technology were unable to stem oil production decline.

The reason for the decline is simple: Oil is a finite resource, which is depleted as it is produced – Production "empties the bucket." When the skeptics tell us that higher prices and improved technologies can reverse oil production declines, recognize that history tells us otherwise. Indeed,

"Oil is a finite resource, which is depleted as it is produced – Production empties the bucket."

> ## "Technology advances often allow us to empty the bucket more rapidly, rather than recover more oil."

technology advances often allow us to "empty the bucket" more rapidly, rather than recover more oil.

E. What happens to exports when production declines?

Countries with oil production that exceed their own needs have exported the excess. But when oil production declines, countries give first priority to their own needs, and exports decline, sometimes rapidly.

Consider the situation in the U.K., which was an oil exporter for over two decades. As shown in Figure V-5, U.K. oil production ramped up dramatically after oil was discovered in the North Sea in the mid 1970s. Production bounced around due to various events and then went into decline around the year 2000. When decline set in, there was no stopping it.

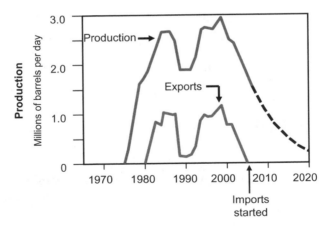

Figure V-5. UK oil production and exports. This history shows that when oil production decline sets in, there is no stopping it, in spite of high oil prices and improved technology. Around 2005 the U.K. became an importer of oil.

> ## "The loss of an exporter means the addition of an importer."

Prior to the buildup of its own oil production, the UK was an oil importer. After production rose above its internal needs, it became an exporter. After its production went far enough into decline, its exports stopped, and it became an importer once again. This example illustrates an important sequence of events: At some point after production in an exporting nation goes into decline, not only do exports stop but imports start and ramp up thereafter. The loss of an exporter means the addition of an importer.

Many countries have transitioned from being exporters to being importers, and more are sure to follow.[6] When dealing with finite resources, this transition is inevitable, and at some point, there simply will not be enough oil production from fewer and fewer exporters to satisfy the needs of others. This is a simple, irrefutable fact.

F. The R/P Trap

From the foregoing, it is clear that oil fields behave in ways that are foreign to experiences in our everyday lives. That fact makes it difficult for most people to get their minds around the realities of oil production. In the foregoing, we have shown that oil production does not stop abruptly. Rather, it declines over a very long period of time, measured in multiple decades. Then why do people talk about years of oil supply?

In many oil producing countries and regions, we have good data on how much oil is produced, and we have reasonable estimates of remaining oil reserves. Reserves are measured in barrels, while production is expressed as barrels per unit of time – Days or years.

Some analysts divide estimated reserves (R) by annual production (P), which yields a number -- R/P in units of years. They then claim that R/P expresses the years that a country or region will have oil production at a given rate of production. The R/P ratio implies a constant rate of production over a future time period, which is not the way oil production behaves.

Consider the situation for the U.S. Lower-48 states, shown in Figure V-6. The data shows that R/P varied over a range from about 14 down to about 8 and then back up towards 14. If the 14 years associated with the 1950 data were adopted as the lifetime of oil production, then production would have ceased by about 1964, which it did not.

> *"The R/P ratio implies a constant rate of production over a future time period, which is not the way oil production behaves."*

People who tell you how many more years of oil production to expect are almost always talking about the ratio R/P. At best, they are uninformed and naïve.

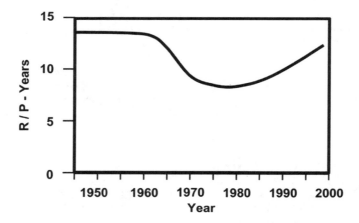

Figure V-6. Reserve to production ratio for the U.S. Lower 48 states for the period 1940- 2000.

VI. World Oil Production, Prices, and World GDP

A. Introduction

This part of the story is also complicated and requires consideration of a range of diverse phenomena and experiences:

- A brief history of world oil production provides some useful perspectives.

- The relationship between oil discovery rates and subsequent oil production illustrates a simple, fundamental fact-of-life.

- Giant oil fields provide 50-60% of world oil production, and their declining discovery rates signal trouble ahead.

- The pattern of recent world oil production shows a troubling trend change.

- World oil production growth and world GDP growth have tracked one another closely for decades. As is often said, oil is the lifeblood of our current economic system, and the data back-up that assertion.

B. Some History

The history of modern world oil production is usually traced back to 1859, when Edwin Drake drilled America's first oil well in western Pennsylvania. Following that seminal event, oil production remained modest by today's standards until around 1900, when many more oil fields were discovered, and there was a rapidly increasing demand for oil products, particularly gasoline for automobiles.

As time went on, more and more capital equipment (cars, trucks, buses, airplanes, boats, power plants, etc.) was built to run on oil-based fuels, and more oil fields were discovered and brought into production. While the history of world oil makes for fascinating reading, our interest here is what lies ahead.

As shown in Figure VI-1, world oil production increased dramatically in line with world economic development over the last century. The trend was interrupted briefly in the mid 1980s, when oil demand declined, largely due to increased oil prices, significant mandated increases in the efficiency of automobiles, and major reductions in oil use in the electric power sector.

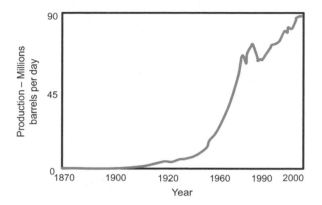

Figure VI-1. World oil production from 1870 to 2009. Measured on the scale of today's production, it's clear that the trend broke upward in the early 1900s and continued unabated, except for the downturn in the mid-1980s when excess supply collided with mitigation resulting from actions taken in the 1970s. Thereafter, the demand increase resumed, generally following increasing world GDP.

Recently, the history of oil prices has been more erratic, as seen in Figure VI-2. Initial high prices were the result of the early scarcity of oil, but within decades, oil prices declined, leveled out, and remained relatively stable for a long period of time, as newly discovered oil fields provided growing oil production in rough lockstep with growing demand.

The 1973 Arab oil embargo abruptly and dramatically changed the world oil market, causing a threefold increase in oil prices. The1979 Iranian revolution led to another oil shortage with another sharp oil price increase. However, in the aftermath of these two shock oil shortages, oil was backed out of electric power production in many places, more efficient automobiles were brought into being, and additional oil production was brought on line from the North Sea, Mexico, and other locations, leading to an excess in oil supply, which led to the price collapse of the mid 1980s.

Prices stayed relatively low until around the year 2000, when increased demand associated with healthy economic growth collided with world oil production limitations, largely because oil discoveries had been declining for decades, as we will discuss.

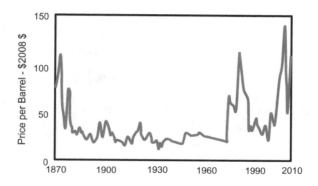

Figure VI-2. World oil prices from 1870 to late 2009 (2008$). Prices were initially very high, when oil first burst into the marketplace. Over time, prices settled down to generally modest fluctuations until the oil embargo of 1973 and the Iranian revolution of 1979. The drop in world oil demand in the mid-1980s, led to a dramatic price collapse, which was reversed in the early 2000s.

> *"Nature created oil in a non-uniform manner around the planet."*

C. Oil Discoveries and Production

Oil production is composed of contributions from a wide array of oil fields, some small, some large, and some very large. Different oil producing countries have varying oil field size distributions, and many countries have no oil production at all. Nature created oil in a non-uniform manner around the planet.

Consider discovery and production rates. Obviously, oil must be discovered before it can be produced. This sounds trivial, but it is useful to remember.

> *"Obviously, oil must be discovered before it can be produced."*

> ## *"It is producing oil fields that provide the oil flows that run our economies."*

Once new oil fields are found, time, effort, and resources are required to bring them into production, particularly giant and supergiant fields, which require huge investments and long periods of time before meaningful production can commence. Again, it is producing oil fields that provide the oil flows that run our economies. Oil in the ground is important, but oil is only useful when it is produced.

To see how some of these factors interact, consider the history of oil discovery and production in the U.S. Lower 48 states -- Figure VI-3. In this case, as in almost all other situations where continuous exploration and development have taken place, the pattern of production following discoveries is similar. For the U.S. Lower 48 states, peak oil production followed roughly 30 years after the peak in oil discoveries, a time lag often repeated elsewhere in the world.[7]

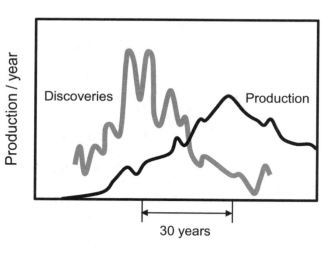

Figure VI-3. The history of discoveries and production in the U.S. Lower 48 states. The peak of oil production occurred roughly 30 years after the peak in discovery.

After the discovery peak, subsequent discovery rates decline, which is not surprising, because there's less and less oil in the ground to be found. In almost all cases, bigger oil fields are found early, because they are typically the easiest to find. Smaller fields are found later, and their discovery often goes on for decades, because there are often many smaller fields to be found.

Keep in mind that finding many more small oil fields does not necessarily result in increased oil production, as illustrated by the U.S. experience.

After oil is discovered, it's almost always immediately developed and brought into production, because of investment realities – Oil companies want returns on their investments as soon as possible. As less and less is discovered, less and less will be brought into production, so production will peak at some point after which it will also decline. It is fundamental: Discovery rates peak and decline, after which production rates peak and decline. The logic is simple.

> *"It is fundamental: Discovery rates peak and decline, after which production rates peak and decline."*

D. Giant fields are extremely important.

In the jargon of the industry, the largest oil fields are referred to as *Giants* or *Super Giants* – "Giants" for short. Smaller fields are sometimes called *Dwarfs*. The reserves associated with these three categories are as follows:

- Super Giants – Reserves above five billion barrels;
- Giants – Reserves between half billion and five billion barrels;
- Dwarfs – Reserves up to half a billion barrels.

As we already noted, billions of barrels are not as large as they might seem when we consider that oil fields typically produce for decades. Divide a billion barrels of oil reserves by 365 days per year and divide again by a highest production lifetime of say 10 years. The conclusion is that daily oil production rate during the best of times is a much smaller number than the stated reserves number.

Worldwide, it has been estimated that there are somewhere between 50,000 and 70,000 oil fields, the overwhelming majority being dwarfs. Giants are responsible for 50-60% of world oil production.[8] About 25% of total world production comes from just 13 giant fields. Of the world's 20 largest giants, 16 are past peak

> *"About 25% of total world production comes from just 13 giant fields. Of the world's 20 largest giants, 16 are past peak production and in decline."*

> *"Since the world consumes roughly 85 million barrels per day, a 10 billion barrel giant field at its peak can contribute a little over 1% of daily consumption during its best years and less later in life."*

production and in decline.[9] This is not a promising trend!

Next, consider the history of giant oil field discovery rates, shown in Figure VI-4. The peak rate of giant oil field discoveries occurred in the 1960s and giant discovery rates have been declining ever since. It is easy to conclude that the discovery-production pattern shown in Figure VI-3 and the giant discovery trends shown in Figure VI-4 portend serious trouble in world oil production in the not too distant future.

Against this background, consider the significance of the recent giant field discoveries in the Gulf of Mexico and offshore Brazil. For the sake of discussion, let us assume that the upside reserves estimates of tens of billions of barrels are correct. Without a doubt, these discoveries are good news. However, their significance on the world scale is modest.

An oil field with reserves of ten billion barrels might produce roughly one million barrels per day at its maximum, after which it will produce less and less each year. Since the world consumes roughly 85 million barrels per day, a 10 billion barrel giant field at its peak can contribute a little over 1% of daily consumption during its best years and less later in life.

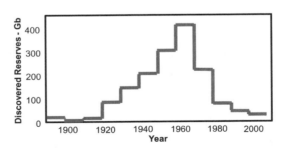

Figure VI-4. Discovery trends for giant oil fields worldwide. Each step represents discoveries for the decade indicated. The peak discovery rate is seen to clearly have occurred in the 1960s.

E. Recent World Oil Production

As noted, the giant oil field discovery rate has been deteriorating for decades, which bodes ominously for the future, since giants are such an important part of total world oil production. From the point of view of world oil supply, we are living on past successes, and we are draining what was found during the best of times.

But the situation gets worse. Consider what has happened to world oil production in recent years. Figure VI-5 shows the situation since 2002. We see that world liquid fuels production flattened out at a mean level of roughly 85 million barrels per day and stayed thereabouts.

> *"We are draining what was found during the best of times."*

At this point, we make an important, practical simplification. Our primary interest is in the production of all liquid hydrocarbon fuels, a large fraction of which are critical to transportation. All liquid fuels include a range of grades of conventional oil and other hydrocarbon liquids. At the beginning of 2010, liquid fuels worldwide were composed of roughly 86% conventional oil and 10% Natural Gas Liquids (NGLs) with the remainder being small amounts of heavy oil and biofuels, which we discuss later. The primary component of total world liquid fuels production is thus conventional oil. As is often done by others, we will refer to all liquid fuels as simply oil. This assumption works well for our general purposes. [10]

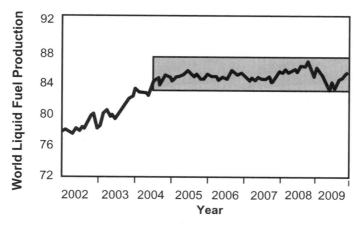

Figure VI-5. World oil production from 2002 - early 2010. World production of all liquids stopped growing in 2004 & has been in a 5% fluctuating band plateau since then.

> *"Oil production data inherently fluctuate, because of a myriad of forces and events that occur throughout the multitude of oil fields around the world."*

In Figure VI-5 we show a band within which oil (all liquid fuels) production has fluctuated. The use of a fluctuation band is a very important concept. Oil production data inherently fluctuate, because of a myriad of forces and events that occur throughout the multitude of oil fields around the world. Included are shutdowns for maintenance, accidents, equipment failures, new fields coming on line, oil fields going into decline, government actions, etc. As a result, to discern a clear breakout trend, we must wait until oil production moves well outside its previous fluctuation band; otherwise we may simply be witnessing a normal or an unusually large fluctuation, rather than a real trend.

As an example of fluctuations that can occur during a plateau, consider European oil production, shown in Figure VI-6. After a relatively stagnant few years during the late 1980s, production rose to a maximum in the mid 1990s.

Thereafter, it fluctuated in a 3% band for about six years. During that period, the data did not allow us to clearly discern what would happen next. In 2002, production began to break downward, but the trend was not verifiable for several years. It was necessary to wait that long to be sure that the break was not just a large negative fluctuation.

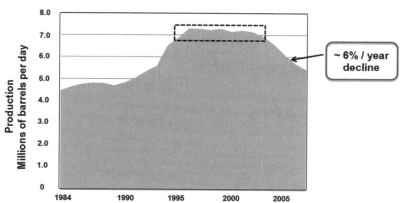

Figure VI-6. European oil production from 1986 - late 2009. Production rose until it hit a fluctuating plateau of roughly 3%, which it maintained for roughly 6 years.

> ## "World oil production has been fluctuating on a plateau for at least six years."

Bottom line: World oil production has been fluctuating on a plateau for at least six years. That production pattern represents a dramatic change from previous production trends. Maintaining the plateau requires major new oil production brought into operation each year, as we explain next.

F. Maintaining plateau production is not easy.

Declines in production occur every year in oil fields around the world, so for world production to remain relatively constant, those declines must be offset by new production.

To illustrate the situation, consider three equal sized oil fields that break into a 10% decline at different times, as shown in Figure VI-7. In this fictitious situation, overall production decline rates increase as each field enters decline. The decline rate in this example goes from 0% to 3.3% to 6.6% to 10% over the period that we considered. On a world scale, the sizes of fields entering decline vary significantly, as does the timing, but the basic pattern is similar to our illustration.

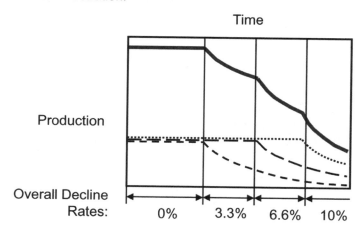

Figure VI-7. Three same-size oil fields that go into declines of 10% per year at different times. As additional fields slide into decline, the overall decline rate increases.

The decline rate for existing world oil production has been a matter of some controversy, but in recent years general agreement has emerged. In

> **"To simply maintain constant world oil production requires adding many millions of barrels per day of new production just to make up for continuing declines in existing oil fields."**

their *2008 World Energy Outlook*, the International Energy Agency (IEA) concluded that the overall world oil production decline rate in existing oil fields is roughly 5%, and IEA expects that rate to increase over time, as more fields enter decline.[11] In other words, the situation will get worse in the future, which is not surprising because more and more older fields will tip into decline as time goes on, while the rate of new large oil discoveries, which lead to future production, continues to decrease. In a recent study, a number of world oil production decline rates from different sources are tabulated.[12] Estimates varied from 2% to 7%, so a nominal rate of 5% is a reasonable number.

At roughly 85 million barrels per day of world oil production and an overall annual decline rate of 5%, the amount of new production needed each year just to maintain level production is roughly 4 million barrels per day, which is a very large amount of new production to bring into production each year.

This is an extremely important point. To simply maintain constant world oil production requires adding many millions of barrels per day of new production just to make up for continuing declines in existing oil fields. In other words, we need to run fast just to stay steady.

Some analysts forecast increasing overall world oil production in the future. For that to occur, more than 4 million barrels per day would have to be brought into production each year. In other words, world oil production would have to increase beyond the recent plateau. We find the likelihood of that happening to be small.

Next consider these circumstances in an economic framework.

> **"Oil prices increased dramatically, largely because increasing demand was chasing stagnant production."**

G. The Relationship Between World GDP and World Oil Production

We have observed that oil is a lifeblood of world economies. To illustrate this point, consider the relationship between world GDP growth and world oil production growth over recent decades, shown in Figure VI-8. (Oil production data is from EIA, April 2007 International Petroleum Monthly, May 8, 2007. GDP Market Exchange Rate data is from the IMF World Economic Outlook Database. April 2007.) It is clear that as world GDP has increased, world oil demand (equal to production) increased in lock step at roughly half the rate.

Next consider what happened since world oil production reached its current plateau. From mid 2004, the onset of the production plateau, to the onset of "The Great Recession" in 2008, world GDP growth was generally in the 4%+ range, as shown in Figure VI-9. So over the period, world economies were expanding -- GDP grew – at a time when there was no oil production growth. The only thing that could change in such a situation was oil price. Figure VI-10 shows that result, based on our "lockstep" reasoning: Oil prices increased dramatically, largely because increasing demand was chasing stagnant production.

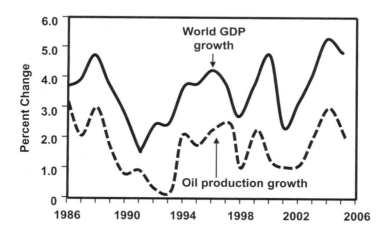

Figure VI-8. Growth in world GDP and world oil production. Over the 20-year period the growth ratio was roughly 2:1, indicating a long history of GDP growth matching oil production growth.

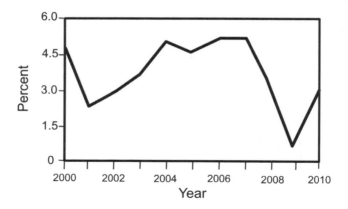

Figure VI-9. World GDP for the period 2000-2010.[13] **Significant economic growth occurred throughout most of the period. Even during "The Great Recession," growth did not go negative.**

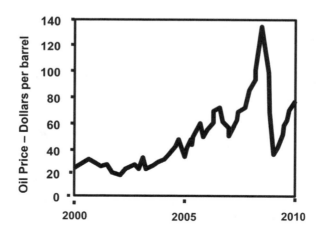

Figure VI-10. World oil prices from 2000 – 2010.[14] **At about the time that world oil production stagnated, oil prices increased dramatically. It was only "The Great Recession" that interrupted the upward escalation. At that, after early 2009, oil prices proceeded to dramatically escalate again.**

There are those who deny the impending onset of the decline of world oil production. Many of them would have expected escalating oil prices over a number of years to have resulted in massive new oil discoveries. That did not happen; rather, new production was only sufficient to balance losses from existing production, resulting in the production plateau.

VII. What Are The Forecasters Forecasting?

A. Introduction – It's not simple to keep score.

To understand oil production forecasts, you must understand the definitions each forecaster is using for the various substances called "oil." Oil is always a mixture of hydrocarbons in liquid form. After that, definitions and categories vary. Terms typically used include oil, liquid fuels, crude oil, petroleum, conventional oil, unconventional oil, natural gas liquids, coal-to liquids, gas-to-liquids, biomass liquids, etc.

One set of definitions used by The National Petroleum Council (NPC) is as follows: [15]

- **Conventional Oil:** Petroleum found in liquid form … flowing naturally or capable of being pumped at reservoir conditions without further processing or dilution.

- **Unconventional Oil:** Heavy oil, very heavy oil, oil sands, and tar sands (bitumen) are all currently considered unconventional oil resources. These compounds have a high viscosity, flow very slowly (if at all) and require processing or dilution to be produced through a well bore.

- **Natural Gas Liquids (NGLs)** are those portions of the hydrocarbon resource that exist in gaseous phase when in natural underground reservoir conditions, but are in a liquid phase at surface conditions, that is at standard temperature and pressure conditions.

These are useful categories but under Unconventional Oil the NPC should have listed Coal-To-Liquids (CTL), Natural Gas-To-Liquids (NGL), and Biomass-To-Liquids (BTL).

Other analysts differentiate oils by where they were found. For instance, they consider "easy-flowing oil" found onshore and in the shallow offshore as conventional, but oils of similar compositions found in the arctic or very deep water are considered unconventional.

A more comprehensive set of definitions is as follows:

1. Crude oil is a high or low quality mineral oil (from the earth) produced anywhere in the world – Onshore and offshore.

2. Natural Gas Liquids (NGLs) are light oils that are produced along with natural gas. NGLs may or may not be included in "crude oil."

3. Oil sands oils are generally those produced in Canada from a tar-like substance. Oil sands oil is usually called unconventional but not always.

4. Synthetic oils are produced by breaking down and rearranging the molecules in coal and natural gas.

5. Biofuels are liquid hydrocarbons produced from plant matter, called biomass (thus "bio"). One example is ethanol produced from corn or other plant material.

6. Shale oil is a mixture of chemical compounds found in sedimentary rocks; in its natural form, it's called kerogen.

Lumping all categories together and calling them "oil" covers everything and that term is what we have used in this book. As we previously noted, the task of identifying trends is simplified by the fact that worldwide conventional "oil" (NPC definition) represents 86% of all liquid fuels produced in early 2010, and Natural Gas Liquids (NGLs) represent an additional 10% for a total of 96% of all liquid fuels.

> **"Lumping all categories together and calling them oil covers everything and that term is what we have used in this book."**

A number of techniques are used to forecast future world oil production, most of them focused on conventional oil, which has been the primary focus of analysts' efforts, because it drives the whole liquid fuels supply situation. All analysts rely on data of differing quality – Some excellent and some highly suspect. The bad news is that analysts sometimes use some so-called "data" that are almost certainly inaccurate, which introduces uncertainties. The good news is that while analysts sometimes use different analytical approaches in formulating their predictions, many come to conclusions that are similar to those developed using other techniques. Precision is not possible in oil production forecasting, but disturbing trends result from a number of different techniques.

> *"Precision is not possible in oil production forecasting, but disturbing trends result from for a number of different techniques."*

Next, we must ask what analysts mean when they use the term "peak oil," which is popularly used. Some consider a small spike in production that may be bigger than other spikes to be "the peak." In our view, a small peak on a fluctuating plateau has relatively little significance.

We consider the term "peak oil" to represent the undeniable onset of the decline of world liquid fuels production, because that point will mark the onset of serious world economic hardship.

> *"We consider the term "peak oil" to represent the undeniable onset of the decline of world liquid fuels production."*

Industry executives often express their opinions about impending oil supply problems in general terms, such as supply/demand "going out of balance," "supply problems" by a certain year, total supply not reaching a certain level, etc. In spite of these and other complications, the results of a number of analyses yield similar results, namely that serious oil supply problems are likely within a matter of years. [16]

> *"The results of a number of analyses yield similar results, namely that serious oil supply problems are likely within a matter of years."*

In Section B, we provide some history of forecasting, a description of a number of the forecasting techniques that analysts have used, and a number of the troubling forecasts that have resulted. In Section C, we summarize a number of organizations that tell us not to worry, that all is well. Understandably, our views are skewed towards deep concern, because of our own analysis of the available studies and evidence.

B. Forecasting Techniques

1. Some History

People have forecast difficulties in oil supply for over 150 years.[17] For instance, in 1874, the state geologist of Pennsylvania, the leading oil-producing state at the time, estimated that only enough U.S. oil remained to keep the nation's kerosene lamps burning for four more years. During World War I, the U.S. government concluded that depleting U.S. oil supplies required reliance on oil-shale resources. In 1952, the Paley Commission forecast that by the 1970s, the United States would have to shift its reliance from oil to coal and synthetic fuels. During the energy crises of the 1970s, a number of studies predicted imminent, permanent oil shortages and oil prices of $100 per barrel in 1980 dollars.

In the earliest years of world oil production, very little was known about petroleum geology, so analysts had no basis for forecasting future production. After World War II, as oil field science and technology advanced and the experience base expanded, more substantive analyses and forecasts became increasingly possible. Over the ensuing years, there were some very insightful studies done, along with some incorrect forecasts.

2. The Hubbert approach

The first widely recognized effort to estimate the peaking of oil production in the U.S. and the world was made in the 1950s by M. King Hubbert, a geologist working at Shell Oil. He reasoned that oil production in a large geographical area would likely follow a bell-shaped curve over time, shown schematically in Figure VII-1. The peak of such a curve would correspond to the peak of production in the region under consideration.

Hubbert made two notable forecasts in 1956. The first was that production peaking in the U.S. Lower 48 states would occur around the year 1970, which is what happened, as indicated in Figure VII-2. The second was that world production would likely peak around the year 2000, which did not happen. He focused on the Lower 48 in part because there was a very good database, and at that time, there was very little oil production in Alaska, since the huge Prudhoe Bay field was not discovered until the late 1960s and it is geologically distant from the Lower 48 states.

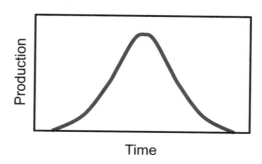

Figure VII-1. Classic bell-shaped curve.

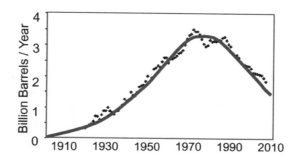

Figure VII-2. U.S. Lower 48 annual oil production. The fit to the classic bell-shaped curve is seen to be remarkably good, considering the fluctuations inherent to the data.

In 1956, relatively little production data was available for the world, and the year 2000 was more than 40 years away. While his world forecast was incorrect, he may have been in the ballpark, if more recent forecasts turn out to be correct.

In later years, other analysts extended the Hubbert method. They observed that by plotting oil production data differently, the data settled down to a reasonably good straight line. The parameter that they plotted was the ratio P/Q, where P was oil production in a given year and Q was the cumulative oil production from the beginning of oil production in the region to the year in question. Q is in units of billions of barrels -- gigabarrels or GB -- measured up to the year in question, and P was annual production in Gb per year. Time does not show on these plots but is easily extracted from the data.

The results using this method for the Lower 48 and the world are shown in Figures VII-3 & 4. In the case of the Lower 48, the straight line is fairly unambiguous. For the case of the world, there is some uncertainty associated with drawing the straight line. Small variations in line selection can lead to tens of billions of barrels of error in the final cumulative oil available. Nevertheless, the trend is clear.

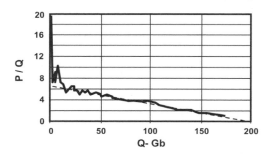

Figure VII-3. Lower 48 oil production data plotted as P/Q versus Q. P is annual production in billions of barrels, called gigabarrels (Gb), per year and Q is in Gb. A good straight line fit is seen after roughly 50 Gb of production. In this case, the peak of Lower 48 production in 1970 occurred at roughly 50% of total estimated production.

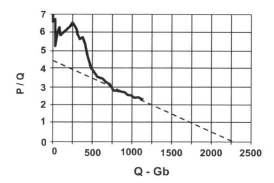

Figure VII-4. Estimated world oil production data plotted as P/Q versus Q. P is annual production in billions of barrels, called gigabarrels (Gb), per year and Q is in Gb. A good straight line fit is seen after roughly 500 Gb of production. In this case, the data in year 2005 was at roughly 48% of the indicated total world production.

The technical literature contains studies detailing the strengths and weaknesses of this approach to forecasting. Nevertheless, whenever data settle down to a straight line, the method and forecasts deserve serious consideration. Understanding the rationale behind the technique is beyond our scope in this book. Figure VII-4 is simply intended to show that the data points fall close to a straight line as time and production increase.

Other Hubbert-based analyses have been performed, some recently. Of particular note was a 2010 study by analysts at Kuwait University.[18] They applied the Hubbert formalism to 47 countries and concluded that world conventional oil production will peak in 2014. The results are striking because of their early peaking forecast and because it is the first peak oil forecast from an OPEC country, breaking decades of denial that peak oil was even worth discussing.

3. Creaming Curves

This interestingly titled technique was developed by Shell in the 1980s to model cumulative discoveries versus the cumulative number of separated exploratory wells, called new field wildcats. The differentiation is between oil wells drilled in new areas to determine if any oil exists at all and other types of wells whose purposes are to delineate a likely discovery or to enhance production.

A description of creaming curves is as follows: [19]

"Conventional wisdom holds that for any given basin or play, a plot of cumulative discovered hydrocarbon volumes versus time or number of wells drilled usually show a steep curve (rapidly increasing volumes) early in the play history and a later plateau or terrace (slowly increasing volumes). Such a plot is called a creaming curve, as early success in a play is thought to inevitably give way to later failure as the play or basin is drilled-up. It is commonly thought that the "cream of the crop" of any play or basin is found early in the drilling history."

The creaming curve technique cannot be blindly utilized. It requires an in-depth understanding of the history of wildcat drilling in the region, considerable geological understanding, etc.

Figure VII-5 shows a creaming curve for world oil discovery. A tendency towards flattening is evident, which would imply that relatively few significant new discoveries lie ahead. For the world, that would imply the impending decline in world oil production. We conclude that the creaming data for the world is suggestive but cannot be taken as a clear predictor.

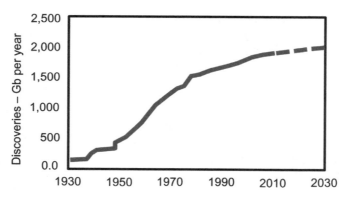

Figure VII-5. Creaming curve for world oil discoveries. The recent flattening of the curve indicates the likelihood of significantly smaller oil discoveries over time.

Another example of a creaming curve is shown in Figure VII-6, which shows the situation in Saudi Arabia. Here the flattening is obvious, which is ominous, since many analysts have been hoping and planning on much greater future Saudi oil production.

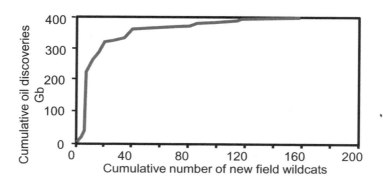

Figure VII-6. Creaming curve for Saudi Arabia. A flattening is evident, indicating that significant quantities of new oil are unlikely to be discovered.

4. Detailed Country Analyses

A number of analysts and the U.S. government have studied individual countries, often in considerable depth. The result is, independent, rigorous country-by-country oil reserves studies and oil production forecasts by a number of geologists and petroleum engineers, many of who are retired from major oil companies and are very experienced.

Of the many credible studies, one is worthy of special note. It was done as part of a doctorate thesis at Uppsala University, Sweden, by Fredrik Robelius under Professor Kjell Aleklett.[20] Robelius built upon the work of others and considered a range of oil reserves estimates for many of the world's largest oil fields. Examples of the range of Ultimate Recoverable Reserves (URRs) he used are shown in Table VII-I.

Table VII-I
Some ultimate recoverable reserves (URRs) used by Robelius

Field name & country	URR range (Gb)
Ghawar (Saudi Arabia)	66 – 150
Greater Burgan (Kuwait)	32 - 75
Safaniya (Saudi Arabia)	21 – 55
Samotlor (Russia)	28
Kirkuk (Iraq)	15 - 25
Cantarell (Mexico)	11 – 20
East Baghdad (Iraq)	11 – 19
Daquing (China)	13 – 18

Some of the ranges are quite large, indicating the range of uncertainties. Use of a range of URRs meant that the study could yield a number of possible world oil production forecasts. Robelius's results are shown in Figure VII-7, where we see a roughly 6-year plateau in world oil production along with three more peaked profiles. To repeat, the dates of greatest interest to us are the dates when world oil production clearly breaks into decline. For these cases, the range is roughly 2011 – 2015.

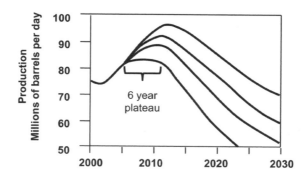

Figure VII-7. Four World oil production forecasts using varying values of ultimate recoverable reserves (URRs) from Robelius.

5. Megaproject Studies

Chris Skrebowski has a long history in the oil business and was formerly the Editor of *Petroleum Review*. A few years ago he observed that a careful examination of the status and outlook for the world's large new oil production projects ought to provide important insights into how much new oil production will come on line in the next decade or so. His approach was based on the fact that large oil production projects take 6-10 years from planning to significant oil production. By following what is in the current construction "pipeline", one can extract a practical view of what new production is likely to be fully operational and when.

To properly perform megaproject estimates, an analyst must understand the histories of similar projects in the regions where the new projects are being undertaken. Oil production capacity growth can be forecast by tracking all the world's 'megaprojects' scheduled to come online and comparing them to known global production decline rates in the world's existing oil fields. Skrebowski's analysis includes deepwater oil, Canadian oil sands, natural gas liquids, and condensate production. His recent conclusion is that world oil production will likely go into decline in 2011 – 2012 period. [21]

After Skrebowski pioneered the megaproject approach, others have performed their own independent analyses and reached similar conclusions.

6. Studies by Organizations

Various organizations have undertaken meaningful world oil production analyses, and a number have concluded that the outlook is ominous. Some notable examples follow.

Over three years ago, the U.S. General Accountability Office did an oil supply study for the U.S. Congress. [22] At the time, the uncertainty range on when world oil decline might begin was much wider than it is today. Nevertheless, their conclusions are of note:

"Most studies estimate that (world) oil production will peak sometime between now and 2040. This range of estimates is wide because the timing of the peak depends on multiple, uncertain factors that will help determine how quickly the oil remaining in the ground is used, including the amount of oil still in the ground; how much of that oil can ultimately be produced given technological, cost, and environmental challenges as well as potentially unfavorable political and investment conditions in some countries where oil is located; and future global demand for oil. Demand for oil will, in turn, be influenced by global economic growth and may be affected by government policies on the environment and climate change and consumer choices about conservation."

"… an imminent peak and sharp decline in oil production could cause a worldwide recession. If the peak is delayed, however, (various) technologies have a greater potential to mitigate the consequences."

The U.S. government response to the GAO report was polite dismissal.

The German Energy Watch Group is composed of independent scientists with oversight by an international advisory board of scientists and parliamentarians. In its 2008 report, entitled "Crude Oil – The Supply Outlook," [23] the group concluded:

- "Peak oil is now."

- "The most important finding is the steep decline of the oil supply after peak. This result – together with the timing of the peak – is obviously in sharp contrast to the projections by the IEA in their reference scenario in the WEO 2006. But the decline is also more pronounced compared with the more moderate projections by ASPO. Yet, this result conforms very well with the recent findings of Robelius in his doctoral thesis. This is all the more remarkable because a different methodology and different data sources have been used."

> **"This is all the more remarkable, because a different methodology and different data sources have been used."**

The Global Witness Foundation is an independent non-profit, public benefit corporation, backed by a foundation that supports studies into "the causes and effects of the exploitation of natural resources by public and private entities throughout the world, specifically where such exploitation is used to fund conflict, human rights abuses and corruption." In its 2009 study, entitled "Heads in the Sand," it observed and concluded: [24]

- "1965 was the year in which the largest volume of oil was discovered. Since then, the trend in the number and average size of discoveries has been in decline."

- "Between 2005 and 2008 conventional oil production ceased to grow, despite massive investment, increasing demand and prices. This failure to increase conventional oil production, despite all the right incentives, is unprecedented in the history of the oil industry."

- "This report has sought to demonstrate that the four underlying fundamentals of oil field depletion, declining discovery rates, insufficient new projects, and increasing demand, though obvious for a long time have not been acknowledged, or acted upon, by virtually any governments with a few notable exceptions. This failure has resulted in a lost decade, with the potential for very serious social and geopolitical consequences from the inability of oil supply to keep up with global demand. In reality there has been a double loss, whereby a proper understanding of the scale and imminence of the oil supply crunch could have injected a sense of urgency to the desperately slow pace and inadequate ambition of governments in the international negotiations to address the climate crisis."

The UK Industry Taskforce on Peak Oil and Energy Security is a group of British companies concerned that threats to energy security are not receiving the attention they merit. In its 2008 report, "The Oil Crunch-Securing the UKs Energy Future", the group concluded: [25]

- "Plentiful and growing supplies of oil have become essential to almost every sector of today's economies. It is easy to see why, when we consider that the energy locked into one barrel of oil is equivalent to that expended by five labourers working 12 hour days nonstop for a year. The agricultural sector perhaps makes the case most starkly: modern food production is oil dependent across the entire value chain from the field to the delivered package."

- "On balance, having reviewed the state of play in global oil production, the taskforce considers that the "descent" scenario is a highly probable global outcome (global production falls steadily as oilfield flows from newer projects fail to replace capacity declines from depletion in older existing fields). We also fear that a "collapse" scenario is possible (the steady fall of the "descent" scenario is steepened appreciably by a serial collapse of production in some – possibly many – of the aged supergiant and giant fields that provide so much global production today). ... In the "collapse" scenario as it might apply to an individual oil-consuming nation, a major oil producing nation - or a group of them - decides that it has been over- optimistic in its assessment of reserves hitherto, that its domestic economic requirements for oil are growing, and it slows or even stops oil supply to nations it formerly exported to. In the

UK's case, the taskforce considers that the "descent" scenario is a highly probable outcome for future UK oil supply. As with the global situation, we also fear that the "collapse" scenario is possible."

In their second report in early 2010, the UK Industry Taskforce on Peak Oil and Energy Security (ITPOES) affirmed their earlier views but concluded that the decline in world oil production is likely to occur in the next five years or sooner. [26] Specifically:

> "The credit crunch of 2008 foreshadowed major economic, political and social upheaval. It stress tested the responses of governments, policy-makers and businesses to the extreme. If only there had been greater time to prepare for its impact and a greater level of understanding about the issues."

> "The next five years will see us face another crunch - the oil crunch. This time, we do have the chance to prepare. The challenge is to use that time well."

Largely because Sir Richard Branson conveyed the ITPOES conclusions directly to the British government, Britain's Energy Minister convened a meeting of business leaders to discuss the government's response to a decline in global oil production, should it actually be imminent. This meeting was the first semi-public meeting by a major world government that accepted the impending decline in world oil production at the outset, rather than beginning by arguing its very existence.

In a 2009 report, the UK Energy Research Centre observed: [27]

- "A global peak is inevitable. The timing is uncertain, but the window is rapidly narrowing.

- "There is a significant risk of a peak before 2020. This is not distant, in view of the lead times to develop alternatives.

- "An increasing number of regions are past their peak of production.

- "The rate of decline of production from existing fields is substantial and accelerating.

- "The timing of the global peak for conventional oil production is relatively insensitive to assumptions about the size of the global resource.

- "The short term future of oil production capacity, to about 2016, is relatively inflexible, because the projects which will raise supply are already committed

- "On the basis of current evidence we suggest that a peak of conventional oil production before 2030 appears likely and there is a significant risk of a peak before 2020. Given the lead times required to both develop substitute fuels and improve energy efficiency, this risk needs to be given serious consideration."

Lastly, in a study by the U.S. Department of Defense Joint Forces Command, "Joint Operating Environment 2008," it was noted: [28]

- "By 2012, surplus oil production capacity could entirely disappear, and as early as 2015, the shortfall in output could reach nearly 10 MBD... The implications for future conflict are ominous... "

7. Oops, the recession!

The recession of 2008 negatively impacted oil production projects to the point that various organizations warned of resulting near future oil supply problems. Here are some sample statements:

- "Oil and gas explorers postponing or scrapping deep water drilling projects are potentially reducing crude supplies by as much as 2.4 million barrels a day in 2011, Morgan Stanley said." [29]

- "The International Energy Agency estimates that about $100 billion of worldwide oil production capacity expansion projects have been cancelled or postponed over the past half year. According to Barclays Capital, oil companies have cut worldwide exploration and production spending by 18 percent so far this year. Deutsche Bank estimated that U.S. energy exploration-and-production spending will drop $22.5 billion this year, a 40-percent, year-on-year decline." [30]

- [Former CIBC World Markets Chief Economist Jeff Rubin] "expects a world oil demand recovery to 86.7 million b/d in 2010, mostly in the developing world, but oil supply at 84.8 million b/d, resulting in a gap of 1.9 million b/d." [31]

- "The International Energy Agency (IEA) ... warns that the credit crisis and project cancellations will lead to no spare crude oil capacity by 2013." [32]

Even the ever-optimistic Saudis issued a warning:

"Saudi Arabia's oil minister warned of a possible "catastrophic" energy supply crunch without prompt investment. ".... such a supply crunch would be catastrophic," Ali Al-Naimi said yesterday. ..... 'The painful result would be felt sooner rather than later. It would effectively take the wheels off an already derailed economy.'" [33]

- Goldman Sachs recently opined: "On our estimates, 2009 will be the last year of growth in non-OPEC production, as the industry suffers from the lack of new project sanctioning in 2007-09." "Our analysis points towards 100% OPEC capacity utilization by 2011/12, leading to the need for demand rationing pricing." [34]

These investment-related concerns add an additional dimension to the finite resource, geological arguments: An investment-related oil production decline could conceivably materialize in a few years. Related oil shortages could be contemporaneous with the geological decline that many forecast to occur relatively soon. If the geological decline was to occur later, then the world could witness one shortage event with a brief recovery, only to be followed by "the big one" a few years later. However events evolve, major oil supply troubles seem unavoidable; exactly how they will unfold is almost impossible to predict, but the outcome is the same: Not good.

> *"An investment-related oil production decline could conceivably materialize in a few years. Related oil shortages could be contemporaneous with the geological decline that many forecast to occur relatively soon."*

8. Some Expert Opinions

A number of knowledgeable individuals have warned of world oil production problems, each expressing themselves somewhat differently. Space limits how many people we can quote. Our selection represents a cross section. We apologize to the legions of other very significant people who are not listed here.

James R Schlesinger was the first Secretary of Energy. [35] At the ASPO meeting in Cork, Ireland in 2007, he stated: "The peakists have won ... to the peakists I say, you can declare victory. You are no longer the beleaguered small minority of voices crying in the wilderness. You are now mainstream. You must learn to take yes for an answer and be gracious in victory." [36]

Matt Simmons is Chairman Emeritus of Simmons & Co. International, and author of the book "Twilight in the Desert," which contended that Saudi Arabia was close to decline in its oil production. He has been sounding the "peak oil" alarm for over five years. Recently, he noted, "Four years have elapsed since global crude output set an all-time record of 73,728,000 b/d. How many added years of falling supply need to happen before we accept that oil peaked?" [37]

T. Boone Pickens, oilman and businessman developed "The Pickens Plan" to solve a number of the U.S. energy problems. When the plan was first issued, it spoke of peak oil as follows: "World oil production peaked in 2005. Despite growing demand and an unprecedented increase in prices, oil production has fallen over the last three years. Oil is getting more expensive to produce, harder to find and there just isn't enough of it to keep up with demand." [38] In November, 2009, Pickens stated that the world has "maxed out" at 85 million barrels a day of production. [39]

Jim Mulva, Chairman of ConocoPhillips, said he expected it will be hard for crude supply to meet demand in the years ahead, with output possibly peaking below 100 million barrels per day. [40]

David O'Reilly, Chevron Corp. Chairman and Chief Executive, warned of a potential oil supply shortfall midway through the next decade that could potentially trigger a substantial increase in prices. ...O'Reilly said there is enough output capacity either on line or coming on line to prevent a supply imbalance in the near term. [41]

John Hess, Chairman and CEO, Hess Corp., at the Oil & Money Conference. Oct. 20, 2009: ".......... growth of production capacity over the next several years will fall short of the incremental 5 million barrels per day each year that we will need to meet demand [meaning to make up for declines in existing world oil production]. We will ultimately be at risk of supply rationing demand through skyrocketing prices that will threaten economic stability and prosperity. If we do not act now, we will have a devastating oil crisis in the next 5-to-10 years."

Charles Maxwell of Weeden & Co. is often called the dean of oil analysts. He recently wrote that peak oil is now projected for approximately 2015. [42]

Christophe de Margerie, CEO of the oil company Total, said that, "If we don't move [now] there will be a problem. In two or three years it will be too late." One result of the underinvestment will be oil prices above $100 a barrel (again), he said. Another result might be "insufficient oil to meet demand" in the years 2010-2015, he said. [43]

Tom Petrie of Bank of America Merrill Lynch has stated that we are at peak oil, and that in spite of BP's "giant" oil find, there will be no change in the oil supply (no surplus). In other words, the world's oil supply will continue to terminally decline. [44]

Fatah Birol, Chief Economist at the International Energy Agency (IEA), in talking about the IEA energy outlook noted that the IEA had done a detailed assessment of more than 800 oil fields in the world, covering three quarters of global reserves. IEA found that most of the biggest fields have already peaked and that the rate of decline in oil production is now running at nearly twice the pace calculated just two years ago. On top of this, there is a problem of chronic under-investment by oil-producing countries, a feature that is set to result in an "oil crunch" within the next five years, which will jeopardise any hope of a recovery from the present global economic recession. [45]

Ray Leonard is CEO of the oil and gas company Hyperdynamics Corp. During his 31-year work in the oil industry, he served in executive positions with Kuwait Energy and Yukos (Russia) after a long tenure with Amoco Corp. He recently opined, "The obvious misunderstanding with peak oil is that the skeptics take a look at the reserves of the world and say—okay, this is how much we're using this year, and this is how much we have, so we'll just divide it by that number for years of reserves. In doing that they completely miss the point that peak oil is about being unable to increase the amount of production when demand continues to increase. Peak oil is not about running out of oil." Peak oil is likely within the next two-to-three years, assuming that demand growth starts to recover. "If this recession continues, the excess supply scenario could stretch out to four to five years." [46]

Irv Miller, U.S. group vice president of environmental and public affairs for Toyota, recently said, "We must address the inevitability of peak oil by developing vehicles powered by alternatives to liquid-oil fuel, as well as new concepts,...." [47]

Steven Chu, the current U.S. Secretary of Energy, recently noted, "I think virtually all Americans are uneasy about our growing dependency on imported oil. I think they are very concerned about looking forward as the energy prices over the long term are expected to increase, simply because production of conventional oil will be peaking, and I think there's some concern there." [48]

Jose Gabrielli, the Chief Executive Officer of Petrobras, the Brazilian oil company, gave a presentation in December 2009 in which he showed world oil capacity, including biofuels, peaking in 2010 due to oil capacity additions from new projects being unable to offset world oil decline rates. Gabrielli stated that the world needs oil volumes the equivalent of one Saudi Arabia every two years to offset future world oil decline rates. [49]

C. Don't worry; all's well.

The other camp in the "peak oil" story includes significant organizations and individuals who tell us "not to worry." Some have an enduring faith that the resource base is very large and that higher prices and improved technology can cure almost any problem.

1. ExxonMobil

Stephen Pryor, president of ExxonMobil Refining in 2006 asserted that "energy resources are adequate to sustain growth— we are not peak oil people." He went on to back up his assertion by saying that the world has thus far produced 1 trillion barrels of oil and that there are still 4 trillion left. [50]

Rex Tillerson, Chairman and CEO of ExxonMobil in March 2009 "celebrated the earth's "continued abundance" of oil, noting that humans have consumed barely a third of the planet's available petroleum reserves. Oil and natural gas, he said, in a story line that Exxon Mobil has perfected over decades, will continue to supply nearly 60 percent of the world's energy needs for the next 20 years." [51]

In a 2006 advertisement, ExxonMobil stated: "Will we soon reach a point when the world's oil supply begins to decline? Yes, according to so-called "peak oil" proponents. They theorize that, since new discoveries have not kept up with the pace of production in recent years, we will soon reach a point when oil production starts going downhill. So goes the theory. The theory does not match reality, however. Oil is a finite resource, but because it is so incredibly large, a peak will not occur this year, next year or for decades to come."

2. BP

Michael C. Daly, then Group Vice President, Exploration & LTR, BP Corp., in a 2007 speech said, "I was asked to talk about oil supply after peak oil. However, I don't accept the premise. It is far from clear to me when there will be a peak to oil supply, at least one driven by a fundamental resource shortage. I believe, from what I know today, that peak oil supply is still a long way off. However, we may face a peak demand for oil first."

David Eyton, BP's Head of Research and Technology, said in Delhi in January 2009: " There has been much debate about if and when we will reach 'peak oil'. BP's viewpoint is that there is no shortage of fossil fuels: we estimate that the world has already demonstrated the commercial viability of around 40 more years of conventional oil resources, 60 years of gas and 130 of coal at current consumption rates. Technology can extend all of these timelines well into the next century, in particular through the development of more unconventional resources."

Christof Rühl, BP's chief economist said, "Physical peak oil, which I have no reason to accept as a valid statement either on theoretical, scientific or ideological grounds, would be insensitive to prices." [52]

3. Cambridge Energy Research Associates (CERA)

Some of the most optimistic analyses of future world oil production have been performed by Cambridge Energy Research Associates (CERA). For many years CERA has assured us that under expected conditions, world oil supply will continue to expand until roughly 2030, after which there will be a long plateau in production. In a recent report it projects "growth of productive capacity through 2030, with no peak evident." [53] They qualified their conclusion as follows:

- There is no unique picture of the course of future of supply; we are dealing with a complex, multicomponent system.

- Aboveground drivers—economics, costs, service sector capability, geopolitics, the timing and nature of government decision-making, and, centrally of course, investment—are crucial to future supply availability.

- Market dynamics will remain highly volatile.

- The upstream oil industry faces major challenges in finding new oil and turning discoveries into commercial production.

We take note of the fact that CERA did not forecast the world oil production plateau that has existed since mid 2004.

Additional statements from CERA:

Robert W. Esser, a geologist and CERA's senior consultant/director of global oil and gas resources, in September 2007 stated, "Peak Oil theory is garbage as far as we're concerned."

Peter Jackson, CERA's Senior Director, Oil Industry Activity, stated in CERA's Multimedia Conference Call, June 20, 2007, "Strong growth in liquids capacity still expected to 2017— 91.0 mbd in 2007 to 112.2 mbd in 2017."

For a number of years, the CERA forecast of world oil production rose above 120 million barrels per day around 2030 and then plateaued, as shown in Figure VII-8. [54] In their 2006 report, CERA stated, "Based on a detailed bottom-up approach, CERA sees no evidence of a peak before 2030. Moreover, global production will eventually follow an undulating plateau for one or more decades before declining slowly. Global resources, including both conventional and unconventional oils, are adequate to support strong production growth and a period on an undulating plateau." .... "Put more simply, the case for the imminent peak is flawed."

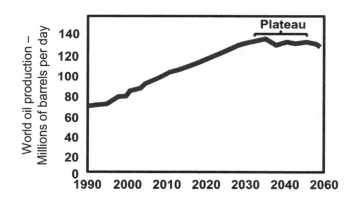

Figure VII-8. Recent CERA world oil production forecast. CERA contended that world oil production would increase above 120 million barrels per day around 2030 and then plateau for more than a decade before declining.

4. Michael Lynch

Michael C. Lynch is President and Director of Global Petroleum Service, Strategic Energy & Economic Consulting, Inc., and one of the better-known resource optimists. He has a long history of challenging "peak oil". For instance:

- "A careful examination of the facts shows that most arguments about peak oil are based on anecdotal information, vague references and ignorance of how the oil industry goes about finding fields and extracting petroleum." [55]

- "The prospect of an oil production peak at 100 mb/d, as some in industry now believe, appears unlikely in my opinion, as most of the above-ground constraints should be overcome. Unless there are serious demand side pressures (which I don't expect), oil production will probably pass 100 mb/d within 12-15 years." [56]

D. What a Major Oil Industry Organization Had to Say

In 2007, a comprehensive energy study was completed by the National Petroleum Council, an advisory committee reporting to the U.S. Secretary of Energy. It was entitled, "Facing the Hard Truths about Energy - A comprehensive view to 2030 of global oil and natural gas," and it covered a very wide range of energy topics. [57] In many ways, it was an objective, credible study that reasonably described important issues, known facts, differences of opinion, and the need for serious U.S. government action. For those willing to devote time and effort, the study makes for an interesting read. Here are a few important passages:

- "Concerns about the reliability of production forecasts and estimates of recoverable oil resources raise questions about future oil supply and deliverability. These concerns are strongly expressed in "peak oil" forecasts in which (1) oil production does not grow significantly beyond current levels and (2) an inevitable decline in oil production is increasingly near at hand. Views about oil supply tend to diverge after 2015, with peak oil forecasts providing the lower bound.

- "Global competition for oil and natural gas will likely intensify as demand grows, as new parties enter the market, as some suppliers seek to exploit their resources for political ends, and as consumers explore new ways to guarantee their sources of supply.

- "Supply forecasts are wide ranging and reflect uncertainty at least partly based on recent difficulty in increasing oil production. Forecast worldwide liquids production in 2030 ranges from less than 80 million to 120 million barrels per day, compared with current daily production of approximately 84 million barrels. The capacity of the oil resource base to sustain growing production rates is uncertain. Several outlooks indicate that increasing oil production may become a significant challenge as early as 2015.

- "Explanations for the variance in projections for both conventional oil and natural gas production are widely discussed as part of the "peak oil" debate. As a result, this study sees the need for a new assessment of the global oil and natural gas endowment and resources to provide more current data for the continuing debate.

- "The energy supply system has taken more than a century to build, requiring huge sustained investment in technology, infrastructure, and other elements of the system. Given the global scale of energy supply, its significance, and the time required for substantive changes, inaction is not an option. Isolated actions are not a solution.

- "There is more uncertainty about the capacity of the oil resource base to sustain growing production rates. The uncertainty is based on (1) the rate and timing at which significant quantities of unconventional oil enter the supply mix, and (2) the ability of the oil industry to overcome growing supply-development risks.

- "The finite nature of the oil endowment and the prospect that production will reach a peak and eventually decline contribute to the debate about oil supply.

- "The underlying decline rate in currently producing fields is not universally well-reported. Many observers think that 80 percent of existing oil production will need to be replaced by 2030—in addition to the volumes required to meet growing demand.

- "Opinions differ about the world's estimated ultimately recoverable oil resource and whether fields can continue to increase production if more than half of today's estimated ultimately recoverable resources (URR) have already been produced."

In spite of these warnings the NPC study received little serious attention in the Department of Energy or the media.

VIII. How is the oil debacle likely to unfold?

Discussion of the threat of world oil production decline -- "peak oil" -- has been in the media for several years. Newspapers and talk shows have covered the issue and done interviews on both sides of the issue; investment houses periodically consider the subject; and governments have commissioned studies, some predicting dire outcomes, as we described.

Media coverage increased in 2008 as oil prices reached ever-increasing highs, and coverage decreased when oil prices declined. The recession became the primary focus of attention. While this waxing and waning is understandable, it does not mask the existing, long term, large-scale depletion of a finite resource that is critical to the fundamental health of modern economies.

In our judgment oil prices have yet to clearly reflect the realities of the impending decline of world oil production; rather, they are driven by the value of the dollar, political events in the Middle East, the economic outlook, weather, inventories, etc.

As we described, powerful and articulate people, various organizations, and OPEC deny that the problem exists, or, if it does, that it will not occur for decades. Their messages range from "not to worry" to "we can manage whatever happens." That their optimistic view is embraced is not hard to understand. Few of us embrace bad news easily. Gasoline prices have been moderate and relatively stable since early 2009. Furthermore, the public and decision-makers have difficulty contemplating a civilization-changing problem that is not yet obvious, especially when struggling with a major, painful and expensive recession, unemployment, foreclosures, and other problems. Besides, decision-makers favor problem clarity, because they then have a clear justification for contemplating serious action. The fact that oil prices dropped so low as the great recession developed seemed to support "not to worry."

As a result of these circumstances, we reluctantly conclude that preemptive action on the world oil supply problem is unlikely until the reality bursts into the public consciousness in a sustained, undeniable manner, at which point it will be too late to avoid very serious consequences. How might that happen? Here are a few possible scenarios:

1. Oil prices might escalate to very high levels as the world emerges from the recession and oil shortages develop. This could hamper economic recovery and cause another recession. When people start hurting financially because of high liquid fuel prices, they will begin to pay serious attention. This scenario is based on increasing public pain that reaches a critical mass at which point the realization of "peak oil" is suddenly widely recognized.

2. A major political leader outside of the U.S. might announce the problem. In so doing, that leader would also announce what their government or company plans to do to mitigate. The recent serious consideration of the problem by the British government might lead to such a proclamation. Such a foreign announcement would almost certainly create a sudden shock and panic, which would spread around the world. In this scenario, the U.S. government would be "caught with its pants down" and would appear to be unprepared.

3. The U.S. president might announce the problem. This seems least likely because recent Administrations have known about the issue and have avoided or denied it for their various reasons. In this scenario, the President would also have to announce significant mitigation plans, which would not be easy, in part because effective mitigation steps will run counter to the current public fixation on global warming, environmental issues, and renewable energy. In this scenario, the U.S. government would seize the initiative and not suffer the embarrassment of being caught unprepared.

We believe that each of these scenarios would create a public shock similar to what occurred as a result of the sudden shortage events of 1973 and 1979. In both situations, the public learned of impending oil shortages with little forewarning with the following results:

- Rapidly spreading public panic and insecurity.
- Rapid fuel shortages induced by people and organizations rushing to top off their gasoline tanks and beginning to hoard.
- A dramatic escalation in prices for oil products, such as gasoline, diesel fuel, heating oil, and jet fuel.
- Employment reductions by businesses.
- Large declines in stock markets, as people rush to minimize their losses.
- The rapid onset of inflation and recession.
- Increasing interest rates along with bond devaluations .

We are well aware that today's economies are different than they were over 30 years ago, but human nature is still much the same, as are the basic laws of economics.

The events of 1973 and 1979 are the world's only real-life experiences with sudden oil shortages. The events that ensued back then are not hard to understand; in fact, they are very human. Against this background, when the reality of world oil production decline becomes obvious, we anticipate sudden, widespread public shock and panic, similar to, but perhaps worse than, what happened during the 1970s.

> *"When the reality of world oil production decline becomes obvious, we anticipate sudden, widespread public shock and panic, similar to, but perhaps worse than, what happened during the 1970s."*

IX. The Realities of the World Oil Enterprise: Don't blame Exxon!

A. Introduction

Before we delve into the physical problems associated with future world oil supplies, it is worthwhile reviewing the structure and realities of the world oil marketplace. A myriad of players are active in world oil markets on an everyday basis. Upstream exploration and production companies deliver oil to intermediate and downstream organizations, including refiners, retailers, transportation companies, and consumers. Buyers and sellers from a number of entities bid for oil every day, establishing price points that can bring more buyers into the market or discourage them. Each day brings buy-sell decisions by many organizations. Tongue–in-cheek, one could say it's a well-oiled machine.

> *"Each day brings buy-sell decisions by a myriad of organizations. "*

Problems can develop when there are disparities between buyers and sellers. While over 110 countries produce oil, the top 10 countries produce over 60 percent of world oil. Add in the next five countries and the top 15 oil-producing countries provide 75 percent of the total. Most of the remaining countries are importers, so the top 15 countries basically control a product that is being consumed by over 200 countries. From a security and price standpoint, this is not an inherently stable situation.

B. OPEC

While many economists have been taught that a cartel was inherently unstable and doomed to failure, the Organization of Petroleum Exporting Countries (OPEC) is an example that has proven otherwise. Economic textbooks have had to be re-written.

OPEC was established in 1961 by five member countries. The organization expanded since, and two countries have withdrawn their membership because their petroleum imports exceeded their exports. Currently, there are 12 OPEC members. A list of the countries and their dates of joining OPEC are as follows:

In Africa:

- Algeria (1969)
- Angola (2007)
- Gabon (1975, withdrew in 1994)
- Libya (1962)
- Nigeria (1971)

In the Middle East and Asia:

- Indonesia (1962, withdrew in 2008)
- Iran (1960)
- Iraq (1960)
- Kuwait (1960)
- Qatar (1961)
- Saudi Arabia (1960)
- United Arab Emirates (1967)

In South America:

- Ecuador (1973, withdrew in 1992, rejoined in 2007)
- Venezuela (1960)

OPEC's primary goal is to protect the individual and collective interests of its members. According to the OPEC statues, it is tasked to do the following:

- Provide a stable international oil price by decreasing harmful and unnecessary fluctuations,
- Provide for the interests of the producing nations by securing a steady income,
- Provide a steady supply of petroleum to consuming countries, and
- Provide a fair return to those invested in the petroleum industry.

The primary mechanism OPEC uses to achieve its goals is to set production quotas for each member country. By adjusting the amount of oil available to the market, OPEC can in-principle significantly influence oil prices and the profits of its member-countries. The idea is simple but in practice, members have not always toed the line. The temptation for a little extra income has been too tempting.

The Arab members of the cartel used their influence over the oil market for their own political purposes in 1973, when they implemented oil embargoes on the U.S. and Western European counties in a show of displeasure over the October 1973 Middle East War. Their actions led to a quadrupling of world oil prices, long lines at gasoline service stations, and what is commonly called, "The 1973-74 Oil Crisis". The U.S. economy was severely affected, and the event highlighted the fact that the economic well-being of developed countries was intimately tied to a steady flow of reasonably and stably priced oil. More importantly, the 1973 action was the first demonstration, since World War II, that the economic security of countries could be significantly damaged by a group that controlled the supply of oil.

The second incident occurred in 1979, when the Iranian revolution led to a dramatic decrease in Iran's oil exports, creating world oil shortages and another world oil price spike. This was a sobering demonstration that an internal conflict in a major oil exporting country could significantly impact oil-importing countries worldwide.

A minor crisis occurred in the lead up to the 1990-91 Gulf War. Just the threat of OPEC involvement in support of Iraq caused oil prices to surge. Observing cause and effect, it became clear to OPEC that they did not want to severely damage the goose laying their golden eggs, providing their income. Disabling oil-consuming economies leads to lower prices and lower profits.

C. The International Oil Companies

From its modest beginning in the 1800s, the petroleum industry grew on a foundation of high-risk-taking investors operating in a capitalist environment. John D. Rockefeller overplayed his hand, and the industry became too concentrated, which led to the breakup of the monopolist Standard Oil Trust in 1911. The result was a number of very large super major oil companies, later called International Oil Companies (IOCs).

In the first half of the 20th century, most of the world's oil production and reserves were controlled by the IOCs, which were vertically-integrated private companies that explored for, produced, and marketed oil products throughout the world. For a long time, the IOCs were known as the "Seven Sisters" -- Standard Oil of New Jersey, Royal Dutch Shell, Anglo-Persian Oil Company, Standard Oil of New York, Standard Oil of California, Gulf Oil, and Texaco. These companies were the dominant force in world oil for a long period of time.

Being well organized and well capitalized in what was an inherently very expensive business, the Seven Sisters, during their heyday, were able to pretty much have their way— worldwide. They had access to most of the world's oil reserves and with their large, dominant position in technology, exploration, production, refining, distribution, and marketing, they benefited from increasing worldwide oil demand, which provided large and growing profits for their shareholders. However, their dominance began to fray in the 1960s after a number of Arab countries began exercising increasing control over their in-country oil production through OPEC, which was formed in 1960.

The situation changed dramatically in the 1970s. The IOCs found their opportunities being eroded, and they subsequently began to merge and acquire each other in response to the changing marketplace:

- Exxon merged with Mobil Oil to become today's ExxonMobil.

- Royal Dutch Shell stayed pretty much the same with minor acquisitions.

- Anglo-Persian Oil Company evolved into BP, which subsequently purchased Amoco and ARCO.

- Standard Oil Co. of New York became Mobil and then merged with Exxon to form ExxonMobil.

- Standard Oil of California became Chevron, which subsequently bought Texaco.

- Gulf Oil was purchased by Chevron with some parts going to BP.

- Texaco became part of Chevron.

The international oil company influence over the oil business continued to decline as more and more oil producing nations decided to take over and manage their own oil operations through their own nationalized oil companies.

D. The National Oil Companies

Today, most major oil producing countries have nationalized their oil interests, and in the process, "disinvited" the international oil companies from their previous positions and contracts. Today, what used to be called "Big Oil" can be considered "Baby Oil" on a production and reserves ownership basis. Yes, the IOCs are very large economic enterprises, but they have been marginalized by nationalization.

National oil companies operate as governmental entities and are agents of national economic, social, and foreign policies. They are operated to further the interests of their host countries, rather than world needs. They typically do not operate with the urgency that is characteristic of the international oil companies, because that is not always in a country's self-interest. Thus, if slower oil development results in larger national incomes because of higher oil prices, and if slower development stretches out their oil reserves, so much the better.

In some cases, national oil company management is highly sophisticated, technologically advanced, and very efficient. The Saudi national oil company

Aramco fits that model very well. At the other end of the spectrum is PDVSA, the Venezuelan national oil company, which is choked, starved, and grossly mismanaged by a dictatorial government.

The phase-over from the dominance of the international oil companies to the current dominance of the national oil companies occurred in small step changes so that most people and governments did not become alarmed. After all, the oil kept flowing and oil prices were generally reasonable until the early part of this century, when prices began to escalate at an alarming rate.

In Table IX-I, we list the original Seven Sisters and show how they merged into today's four survivors. We also list today's largest oil companies, which have been called the "Big Seven." [58]

Another way of viewing the shift of control is shown in Figure IX-1, where the access to the world's oil and gas in 1970 is compared to the situation in 2009. [59] In 1970 the International Oil Companies had access to roughly 85%. Today, the surviving companies have full access to roughly 8% along with some access to another 12%. The shift in control is indeed stark. For those who believe that investor-owned oil companies provide the most reliable, secure sources of oil, the current situation is a source of unease, if not alarm. Of the current largest world oil companies on the list, some entities have proven themselves reliable and relatively non-political in recent years, while some of the others are tools of governments that do not have the free world's best interests at heart and could create chaos, if their host governments so dictated.

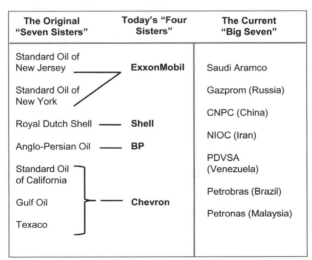

The Original "Seven Sisters"	Today's "Four Sisters"	The Current "Big Seven"
Standard Oil of New Jersey	ExxonMobil	Saudi Aramco
Standard Oil of New York		Gazprom (Russia)
		CNPC (China)
Royal Dutch Shell	Shell	NIOC (Iran)
Anglo-Persian Oil	BP	
Standard Oil of California		PDVSA (Venezuela)
Gulf Oil	Chevron	Petrobras (Brazil)
Texaco		Petronas (Malaysia)

Table IX-I. The former "Seven Sisters", the survivors and the current largest.

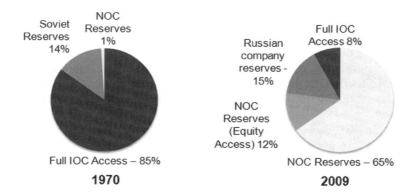

Figure IX-1. Access to the world's oil and gas in 1970 compared to the situation in 2009.

Looking at the differences in the characteristics of privately held International Oil Companies (IOCs) and National Oil Companies (NOCs), we see that privately held companies:

- Maximize value and profit for shareholders through higher rates of return, minimizing costs, and constantly seeking efficiencies.

- Are very mindful of the need for reserve replacement and growth with an emphasis on timely exploration and development.

- Respond to market signals by speeding-up or slowing-down exploration, development, and production.

In contrast, many NOCs:

- Make decisions based on host country objectives, including maximizing revenues for the national government, often at the cost of needed oil field operations.

- Divert substantial portions of their revenues to government programs rather than to reinvestment.

- Support in-country subsidies on petroleum products.

- Are subject to national control on what supply levels and price for petroleum products are available to citizens, the government, and the industrial economy, effectively manipulating industrial structure.

- Can be hugely labor intensive and inefficient, acting as jobs programs within the country.

- Are subject to government controls that dictate supply and price to foreign countries based on government foreign policies.

- Are guided by government decisions on global strategic and energy security goals.

- Have a short investment time horizon in order to produce revenues for the government's current accounts, often sacrificing investments to expand production and reserves.

- Have to compete for capital funding from the government for projects rather than have the decisions made based on financial return or investment standards.

- Must suspend or cancel partnership agreements based on government fiat.

- Do not devote significant revenues to research and development.

These represent some of the major issues that can distort an NOC's business decision-making process. A few NOCs have governments that encourage them to operate with private-sector objectives in the competitive global oil commodity market. However, the majority do not.

Recently, because the IOCs offer high technology knowledge, efficient management, and additional capital, a number of countries NOCs have requested that the international oil companies bid on various projects. But the contract conditions are often onerous. Some countries are 1) asking that 50-80 percent of the profits go back to the host country through royalty fees and taxes, 2) that technology be shared with the NOC, and in some instances, 3) that the equipment required for the project be made in the country. Even after negotiating acceptable terms, the governments may step-in and disinvite the international oil company at any time. This approach to supplying the world's oil has serious faults. Nevertheless, it is the situation that currently exists in many places.

E. Conclusions

As we will show later in this book, the uninterrupted supply of moderately priced oil is essential to world economic well being, let alone growth. Besides the geological threat that lies ahead, which we will describe, the world needs well functioning, stable, and dependable organizations to find, develop and deliver that oil. Based on the evolution and current state of the world oil enterprise, the organizational prognosis for the future is not good.

Finally, consider Exxon. In the past when oil prices escalated and people felt pain, one of the primary scapegoats was Exxon. It was and is very big and has made huge profits. Because it markets in so many locations as Exxon or Esso, its gasoline prices were and are publicly displayed for people to easily see. Independent of any real faults or failures, it became the scapegoat for oil-related pain and suffering.

When world oil production decline occurs and people again want a scapegoat, be mindful that the primary oil scapegoat in the past – Exxon -- no long qualifies. As we have shown, Exxon is now a minor player in the oil world. While people may reflexively trot out the old scapegoat, honest people will realize that they will have to identify a new one.

Exxon provides an indirect measure of the scale of the world oil industry. Exxon is a huge enterprise but only a minor player in world oil and serves to demonstrate just how enormous the world oil industry actually is.

X. Administrative Mitigation

A. Introduction

When world oil production goes into decline, there will be annually growing oil shortages and much higher prices for oil products, e.g., gasoline, diesel fuel, jet fuel, home heating oil, various chemicals, lubricants, etc. People, companies, and governments will attempt to do everything in their power to mitigate the myriad problems that ensue.

We consider two classes of mitigation. The first is Administrative Mitigation, which covers options that can be implemented by individuals, organizations, and governments, either voluntarily or via mandates. The second category is Physical Mitigation, which includes construction and deployment of new vehicles, equipment, machinery, and liquid fuel production techniques and production plants.

In this chapter, we discuss the following administrative options: Government rationing of oil and oil products, car pooling, and telecommuting. In the next chapter, we discuss physical mitigation.

B. Rationing Options – A Rat's Nest of Complications!

1. Introduction

When oil shortages develop and prices escalate, government intervention of some sort will likely happen or be required. Rationing to moderate the effects of the oil shortfall is an important but extremely complex undertaking.

Here we analyze four generic oil and oil product allocation options that could be implemented: (i) oil price and allocation controls, (ii) coupon gasoline rationing, (iii) a variable gasoline tax and rebate, and (iv) no oil price controls with partial rebates. The choice of appropriate policies is by no means obvious, as we will show. Furthermore, their implementation is difficult and imprecise, and they cannot eliminate negative macroeconomic effects. The discussion here focuses on the U.S. but is applicable to most developed nations.

We have tried to make the discussion that follows as straightforward and as readable as possible. Nevertheless, we realize that many readers will have difficulty following the complexity, intricacies, and unintended consequences of these approaches to the non-market allocation of oil shortfalls. The onset of the decline in world oil production will cause widespread economic hardship and social unrest and will undoubtedly be accompanied by demands that governments "do something" to remedy the situation. Rationing is certain to be an option on the table. The following discussion indicates that government rationing and redistribution of oil products is inherently complex and controversial and raises its own set of problems, costs, and inequities.

> *"The onset of the decline in world oil production will cause widespread economic hardship and social unrest and will undoubtedly be accompanied by demands that governments "do something" to remedy the situation."*

As in previous oil shortages, unless the government imposes price controls on domestic oil production, oil prices within a country will rise with the world price, and the increases will be passed on to consumers in the form of higher prices for oil products. For the shortfalls likely to occur when world oil production goes into decline, the size of oil price increases is impossible to estimate, but increases of 100 to 300 percent are certainly possible.

Oil shortages will reduce a nation's production of goods and services directly. Industries will be forced to reduce their oil use, resulting in lower output and higher prices for their products. Any approach that allocates a shortfall of oil cannot offset the supply-side costs, although efficient allocation could minimize them. Further, as workers seek to protect their real incomes, higher prices for oil products could be reflected in demands for higher wages, setting off a wage-price spiral that would compound the inflationary effect of higher oil product prices. Outlays for programs indexed to price levels (such as cost-of-living contracts and pensions) would rise automatically.

2. Criteria for Comparison

There are several criteria for comparing oil-rationing approaches; we have chosen the following four:

> *"Oil shortages will reduce a nation's production of goods and services directly."*

1) **Microeconomic effects**. Each approach will affect the prices of goods and services, inventory management of manufactured goods, and inter-industry and inter-regional relationships.

2) **Macroeconomic effects**. Each approach will affect national economic growth, inflation, and other important macroeconomic variables.

3) **Equity**. Some parts of national economies will benefit and some will lose. People in similar situations may not be treated equally. Rationing approaches may be perceived as fair or unfair across different income groups and regions of a country.

4) **Practical problems**. This includes all the costs of implementing each approach. For example, regulations to prevent fraud and abuse may be required with an allocation approach that distributes gasoline coupons or cash. What impacts on each approach will lobbying activities have? Will politically powerful constituencies develop and influence implementation?

3. The Workings of Each Approach

Price and allocation controls. This approach could impose domestic oil price controls, an entitlements program, and regulations similar to those that were in effect in the U.S. from 1973 through January 1981. Price ceilings were imposed on domestic oil production so that prices would not rise with world oil prices. Since some refiners had access to price-controlled domestic oil while others were forced to buy world-priced imported oil, the government imposed an entitlements system, averaging the prices of domestic oil and imported oil, so all U.S. refiners paid approximately the same price. Thus, even under price controls, the average price of crude oil rose. Controls on the price markups of downstream operators (crude oil resellers, refiners, wholesalers, and retailers) were also required; otherwise, they would be able to raise prices to market-clearing levels and thus negate the effect of price controls.

Allocation regulations were imposed in the U.S. under which suppliers were obligated to sell proportionately reduced volumes to their historical purchasers. Thus, if a wholesaler were able to satisfy only 80 percent of its demand for residual fuel oil, the supplier reduced deliveries to all historical residual fuel oil customers by 20 percent. The government designated priority users, who received a higher portion. For gasoline and diesel fuel, however, the allocation system stopped short of retail purchasers; that is, consumers were not allocated specific quantities. Each gasoline station

> ## *"Even under price controls, the average price of crude oil rose."*

received reduced supplies of gasoline based on historical purchases, so the demand for gasoline at the controlled price was greater than the quantity supplied, so cars waiting in gasoline lines resulted (Figure X-1).

Further actions were required when refiners produced less of a given oil product than deemed desirable. This could occur in the case of home heating oil, and the government had to direct refiners, through refinery yield orders, to change their relative yield of products -- for example, to produce less gasoline and more fuel oil.

Coupon gasoline rationing. This option would impose coupon rationing for gasoline allocation controls, but it would retain the other controls described above. It would limit price increases for gasoline, as would the first option, but would use ration coupons rather than queues to allocate gasoline. Coupons could be distributed on the basis of registered vehicle ownership, on the basis of historical use, and to priority and hardship users. Since coupons could be sold, market forces would set coupon prices.

Figure X-1. U.S. gasoline lines during the "Oil Crisis of 1973." [60]

Price controls would ensure that the price of gasoline does not rise to market levels. Gasoline price controls do not set price ceilings but instead control profit margins, so price controls on domestic crude oil would be required to keep the average refiner acquisition costs of oil from rising as quickly as world oil prices. Without domestic price controls and margin controls on downstream operations, gasoline prices would rise to market levels, causing ration coupons to decrease in value over time and eliminating the need for rationing. Without price and allocation controls on oil products other than gasoline, refiners could increase revenues by increasing the prices and production of these products. A system of local boards would be established to administer state ration reserves, providing additional allotments to those who would otherwise experience severe hardships. Ration coupons would be required for the purchase of gasoline and would then be transferred from retailers up the distribution chain to refiners and finally back to the government.

Gasoline tax and rebate. This approach would consist of a system of emergency gasoline taxes and rebates structured to have effects similar to gasoline rationing with a free market in coupons. Price and allocation controls on gasoline would be prohibited, but they would be required on oil and oil products to keep refiners' oil costs from rising to world levels and to prevent downstream margins from rising. The level of the variable tax on gasoline would be set so that refiners could pass through increases in average oil prices. Thus, as world oil prices and gasoline demand changed, the size of the tax would change. Tax revenues would be rebated to registered motor vehicle owners, just as coupons would be distributed under rationing.

> *"A system of local boards would be established to administer state ration reserves, providing additional allotments to those who would otherwise experience severe hardships."*

A specific example illustrates the similarities between the two approaches. Under the rationing approach, the government would control the gasoline price at, say, $5 a gallon. A ration check (or electronic transfer) for a set number of ration coupons, each good for 1 gallon, would be sent to each vehicle registrant, with a limit of three registered vehicles per household. Assume that each car owner would get 10 coupons a week. Persons requiring more than 10 gallons a week could purchase coupons from those needing less than 10 gallons. As a free market for ration coupons develops, the going price might settle at about $6 a coupon under a 20 percent oil shortage. Thus a gallon of gasoline could be purchased for $5 and a coupon worth $6. The market-clearing price of gasoline would be $11 a gallon -- $5 for a gallon of gasoline plus $6 for the coupon.

Under the gasoline tax and rebate approach, the results would be similar. For a shortage of the same size, the market-clearing price would still be $11 a gallon. An emergency gasoline tax of $6 a gallon would add $6 a gallon to the oil distribution chain. The tax and rebate approach would

> *"The market-clearing price of gasoline would be $11 a gallon -- $5 for a gallon of gasoline plus $6 for the coupon."*

permit the oil distribution network the same profits as the coupon rationing approach. Instead of providing 10 coupons worth $6 each, the government would send $60 a week to each vehicle registrant, with (perhaps) a limit of three vehicles per household.

General rebate. Here the prices of all oil products would be allowed to increase. Prices would determine how much of each product would be produced and how the products would be distributed, and domestic oil prices would be allowed to rise. A windfall profits tax would capture a large portion of the higher oil revenue, which would then be rebated. Rather than trying to avoid price increases by using price controls, this option accepts the price increases and attempts to offset the negative economic consequences with rebates. Whereas gasoline consumers would be most directly affected under the preceding two approaches, all consumers would be affected under the general rebate approach, which would presumably distribute the rebates more broadly. The distribution mechanism could take many forms, such as adjustment in federal income tax rates and changes in withholding liabilities, changes in existing transfer payments, and reduction in the federal debt. It is worth noting that "windfall profits" taxes were implemented in the U.S. between 1980 and 1987, and the results were hugely counterproductive, according to a 1990 Congressional Research Service report. "The WPT reduced domestic oil production between 3 and 6 percent, and increased oil imports from between 8 and 16 percent," says the report. "This made the U.S. more dependent upon imported oil." [61] It took many years after the tax was terminated to settle all of the issues that the U.S. windfall profits tax created.

The recessionary effects of a liquid fuels shortage will reduce tax receipts in the non-energy sectors of the economy. Reduced revenues, combined with increased unemployment and other transfer payments, will create budgetary demands that could be financed by the increased income from the windfall profits tax. Some portion of the rebate could also be used to assist firms by reducing corporate income taxes; businesses, particularly oil-intensive ones, would have to raise the prices of their products as oil

> *"All consumers would be affected under the general rebate approach, which would presumably distribute the rebates more broadly."*

prices rise. Because consumers would receive income supplements in the form of rebates, most businesses would be able to maintain their market. Some, however, would not. For example, California fruit growers who rely on truckers to transport their produce to East Coast markets would find their goods less competitive with fruits produced closer to the market.

4. Microeconomic Effects

Price and allocation controls. Under this approach, prices would not rise to market levels and queuing would result, because of oil product shortages. However, the true price of gasoline, including the cost of waiting in line, would rise to market-clearing levels, and the "waiting cost" would reduce consumer welfare just as would a gasoline price increase. The difference is that without price controls, a higher dollar price is paid to others in the domestic economy and does not involve a loss in well-being for a nation as a whole. An increase in the effective price of gasoline to consumers produced by queuing is a net loss to society, a loss that could be equivalent to hundreds of billions of dollars in the U.S., for example. Further, to the extent that government-determined allocations diverged from allocation to highest value uses, significant losses in national economic efficiency would also be incurred. An allocation rule based on historical use would not be able to keep up with changing patterns of demand or determine which customers could reduce consumption most efficiently.[62] In addition, government officials would be required to decide the priorities of oil distribution, and interest groups would likely influence these decisions.

> **"An increase in the effective price of gasoline to consumers produced by queuing is a net loss to society."**

Inventory behavior would be directly affected by the allocation approach selected. To the extent that future profits from storing oil would be limited by price controls, rationing, or taxes, less oil might be stored by the private market prior to the disruption. During the shortfall, price controls would delay oil price increases; thus it would be in the inventory holders' interest to hold stocks while the price increases. Without controls, however, the unconstrained price would increase rapidly. After reaching the market levels, prices may stabilize, reducing the incentive to maintain high oil inventory levels.

Coupon gasoline rationing. Coupon rationing would eliminate gasoline lines and allocate gasoline supplies to government specified highest value uses. However, this approach would introduce inefficiencies into the oil market. If the bulk of shortfalls were borne by gasoline, the approach would

> "Coupon rationing would eliminate gasoline lines and allocate gasoline supplies to government specified highest value uses."

allocate other, underpriced refined petroleum products to those who might otherwise conserve. Very expensive conservation measures might be forced on gasoline consumers, while relatively inexpensive conservation efforts for other petroleum products would be forgone. For example, commuting by automobile might be made prohibitively expensive, especially for low-income persons, rural households, and those with very long commutes, and industries that depend on automobile traffic would be severely affected. In contrast, if the shortfall were distributed across all oil products, all consumers would find more efficient ways of conserving petroleum products, eliminating the need for gasoline consumers to take extreme measures. A system of price and allocation controls would be imposed, resulting in the inefficiencies described earlier. Finally, no incentives to increase domestic production would be provided by gasoline coupon rationing.

Gasoline tax and rebate. In two respects, this approach is similar to gasoline rationing: (i) the losses incurred if the shortfall were borne primarily by gasoline would be the same, and (ii) the losses from imposition of price controls would be the same. In both cases consumers do not face market-determined prices for gasoline, and incentives to increase domestic oil production are absent. However, because the tax and rebate approach would not require gasoline price controls, it could result in a more efficient allocation of gasoline supplies. The delay between purchases and rebates might impose hardships on many consumers.

General rebate. The general rebate approach would minimize micro-efficiency losses. By encouraging conservation in the use of all refined products, it would put the available supply of oil to its highest valued uses, and it would avoid socially divisive queues for gasoline.

5. Macroeconomic Effects

Price and allocation controls. Like the other options, this approach would not mitigate the supply-side macroeconomic costs of the oil shortfall. The higher effective price of oil would reduce real GDP and raise the general price level. The costs of queuing up for gasoline would be enormous, though not measured directly. For example, in the U.S. in the second quarter of 1979, real GDP decreased 2.3 percent at an annual rate, and much of this decrease was attributed to the oil shortfall. [63] On the demand side, however, oil price ceilings, if effective, would limit the transfer of funds to oil producers and the resulting oil price drag on non-oil markets. Ceilings on oil product prices would also limit increases in the consumer price index (CPI).

Coupon gasoline rationing. This option would not alter the supply-side effects of an oil shortfall, but by controlling domestic oil prices, it would tend to limit fiscal drag.[64] In contrast to a gasoline tax and rebate approach, the ration coupons (paper or electronic) provide a second currency that may insulate some sectors of the economy from fiscal drag. In addition, if the coupon price were excluded from the CPI, the inflationary impact could be reduced. The microeconomic efficiency benefits over the price and allocation controls discussed above could result in less reduction in economic activity, since the rationing approach permits free-market trading of coupons and would allocate gasoline more efficiently than would direct allocations. As a result, Gross Domestic Product (GDP) would tend to be higher and the price level lower than would occur under price and allocation controls alone. On the other hand, if the shortfall were borne primarily by gasoline consumers, economic activity would be retarded, relative to the situation with no controls.

> ## "A rationing system would, in effect, create a new currency."

Gasoline tax and rebate. This approach cannot alter the supply-side effects of oil shortages. A tax-rebate system that includes price controls could limit oil price drag, [65] provided the tax is rebated simultaneously. In the more likely event of uncertain and uneven rebates, the adverse impacts could be sizable. Further, the CPI would directly reflect the gasoline price increase, and this effect would be significant. For example, under the conservative assumption of a doubling of gasoline prices, the CPI could increase by more than five percent in the first month of the program (an 80 percent annualized rate). The increase in consumer prices would trigger increases in indexed wages and entitlement payments. More rapid wage inflation would increase production costs in the economy, and the inflationary impact would be prolonged by second- and third-round effects on wages and prices.

General rebate. This approach would not mitigate the supply-side costs of the shortfall, but, by allowing all supplies and demands to interact at market-clearing prices, decontrol would achieve a greater degree of economy-wide efficiency and a higher GDP than any of the alternative approaches. Even if rebates were distributed immediately, some oil price drag would occur, and funds would flow from non-oil to oil sectors. In addition, rising oil prices would increase the CPI. However, the reduction in economic activity must be balanced against the efficiency gains in terms of both resource allocation and administrative costs of a decontrol system.

6. Equity

Price and allocation controls. Under price and allocation controls with queuing, there is a transfer of income from those who value their time more than the average, to those who value it less or have no choice. This transfer of income may be monetized as the size of the queues increase, as persons with a high value of time may pay others with a low value of time to wait in line. Further, it is not clear that historical allocation is "fair," because regions of the country that are growing more rapidly than others would probably feel they were being treated unfairly. Because of the shift in consumption and driving patterns over time, allocation on a historical basis would become more and more unfair. [66]

Coupon gasoline rationing. The fairness of this proposal obviously depends on the distribution of the coupons. If fairness means reestablishing an individual's purchasing power prior to the disruption, a rationing approach should distribute the coupons according to the amount of gasoline consumed prior to the disruption. Alternatively, if it means providing equal assistance to all income groups, or more assistance to lower income groups, coupon distribution in proportion to automobile ownership may not be appropriate. In the extreme, coupon rationing could be used explicitly to distribute income to lower income groups or to those with special needs.

Gasoline tax and rebate. If the rebate were allocated in the same way as coupons, the distribution of income would be similar to that under gasoline rationing. However, the distribution of money would make the implicit redistribution of income more evident to the public. Distributing coupons to vehicle owners may be perceived as fair, whereas the equivalent distribution of money may not. There is also the question of when the rebates are provided. If infrequently, there will be cash flow problems for the less well off.

General rebate. Uncontrolled prices would be perceived by the public as inequitable, and money would be openly transferred from consumers to oil companies. A general rebate would only partially compensate for this transfer, although the size of the rebate could be increased by adding an emergency surcharge to a windfall profits tax, which is inherent to this approach. If the rebate is distributed to all citizens, some groups will argue that others are receiving too much. Thus, in terms of the distribution of income, the fairness of this program depends on the tax rate and the structure of the rebate mechanism.

7. Practical Problems

Price and allocation controls. A major practical problem with this approach is that a comprehensive price control and allocation system would have to be imposed. Each oil company would be required to submit detailed information to federal agencies, and each refiner would have to report all oil purchases so that the government could determine the net entitlement obligations for each refiner. Oil companies would be required to maintain records of all transactions to enable government agencies to conduct audits, which would require a large federal workforce. Initiating the program would take at least several months, and even if a rationing program was developed and kept intact in standby status, changing circumstances could make any control system obsolete.

Gasoline queues have proved to be socially divisive in relatively small oil shortfalls, but may be less divisive in a clear national emergency. Price and allocation controls without a system of end-use allocation would result in very long lines, which would require increased security costs. Further, once in place, these controls may not be easily removed. The U.S. history of oil regulations might be a guide: The emergency controls program enacted in 1971 and 1973 did not end until 1981. In the future, as in the past, many consumers and oil companies might oppose removal of price and allocation controls after shortages ease.

Domestic price controls would allow oil-exporting nations to raise prices without the reduction in demand that would otherwise accompany such a price increase. Under price and allocation controls, the true price to the consumer is the sum of the controlled price and one of the following: The cost of waiting in line, the size of the tax, or the value of the coupon. Since the real price determines consumption, oil-exporting nations could increase the price, thus reducing the length of the lines, the size of the tax, or the value of the coupon, without affecting the quantity of oil consumed in the U.S. Without controls, however, a price increase would raise the true cost of oil and result in a reduction in demand.

Coupon gasoline rationing. A major problem with this approach is that an oil price and allocation controls system would have to be imposed. Moreover, a rationing system would, in effect, create a new currency and would entail the creation and operation of a massive system parallel to the monetary system to create, disperse, transfer, and, eventually, return coupons or electronic credits to the government. On the other hand, price controls may significantly simplify the task of managing monetary and fiscal policies.

Much of the administrative cost and time required in the preparation and operation of rationing would stem from the employment and training of large numbers of government workers. Costs to the private sector would also be high. Another problem is that the information in the national motor vehicle registration file would be obsolete; the error rate could be very significant. Millions of U.S. automobiles change ownership each year, and

> *"Much of the administrative cost and time required in the preparation and operation of rationing would stem from the employment and training of large numbers of government workers."*

coupons or electronic credits might be sent to previous owners of used cars while current owners receive none.

Gasoline tax and rebate. A major practical problem would be the imposition of a price control system and, in addition, a system of gasoline taxes and rebates would have to be legislated and implemented. Because money would be used instead of coupons or credits, many existing transfer mechanisms might be used; for example, the existing excise tax on gasoline could be raised to the desired level. Some major adjustments may have to be made to control inventory profits. However, only with considerable effort and public and private expense could existing government mechanisms such as income tax withholding, veterans' benefits, low-income energy assistance, assistance payments, or other methods be used to distribute the rebates. If rebates were distributed on the basis of motor vehicle ownership, information on motor vehicle registrations would have to be maintained. If they were distributed by adjusting income tax withholding rates, tax credits might be required for automobile owners whose rebates exceeded their tax liability. Procedures to deal with non-taxpaying or unemployed vehicle owners would be required. Because of the large income transfers, strong incentives to cheat would exist. Additional federal employees would therefore be required to monitor compliance.

An excise tax could be set at a level that equates supply and demand; however, setting the tax at such a level is a very difficult task, and the macroeconomic consequences of mistakes could be severe. It is not clear that the government would be able to determine the correct price, for there are no precise indicators of market equilibrium, which is likely to change as shortages increase year after year. After setting the initial tax, the government would have to adjust the tax on a weekly or monthly basis as crude oil supplies and prices changed and as demand became more elastic with time. The inherent uncertainty in estimating the actual level of the tax would make accommodating fiscal and monetary policy very difficult. Further, this system would be difficult to dismantle. Rebate recipients who used less gasoline than average would not want to give up their rebates, and if history is any guide, price controls would not be easily removed.

> *"Strong incentives to cheat would exist."*

General rebate. Very large interruptions may strain the ability of market mechanisms to function effectively. On the other hand, a major advantage of this approach is that price and allocation regulations would not have to be imposed. No new tax mechanism would be required, but new rebate mechanisms would be needed, especially to handle the enormous revenues generated by large disruptions. With considerable effort the rebates could possibly be handled as an increment to existing programs -- for example, through increased transfer payments and refundable income tax credits. However, if rebates were made strictly per capita, the rebate mechanism would be even more difficult. Assembling a master list of all citizens for the purpose of distributing rebates might be construed as an unprecedented invasion of privacy, especially since any approach would have to be brought to a state of readiness in advance of an oil emergency. If the income tax system were used, many people who do not now file income tax returns would need to do so to receive the rebate. If the oil shortage was large enough to drive prices up to the level at which the rebate would be greater than many families' withholding liability, refundable credits or a combination of withholding reductions and sales or payroll tax reductions may be required.

> *"Assembling a master list of all citizens for the purpose of distributing rebates might be construed as an unprecedented invasion of privacy."*

Procedures would have to be established to deal with hardship cases, exceptional needs of medical patients, low-income users of fuel oil, and related cases. However, the rebate program should be designed carefully to avoid measures that encourage oil use; for example, it should not reward homeowners who continue to consume home-heating oil at pre-shortage levels. All oil should be priced at its replacement value; the rebate should assist those in jeopardy of suffering and should generally restore purchasing power to the economy. The time required to implement an emergency allocation approach is also important. If, as expected, world oil prices rise rapidly at the outset of the decline in world oil production, approaches that are difficult and time-consuming to implement may be less useful; demand reductions under the general rebate approach would be immediate.

8. Overall Evaluation

The rebate approach allows individual firms and consumers the most flexibility in adapting to the oil shortfall and provides the greatest incentive for increased domestic oil production and storage. The coupon rationing and tax-rebate approaches allow efficient allocation of gasoline, assuming the government does its job perfectly. If the burden of the oil shortfall is placed primarily on gasoline, these two approaches are less efficient in microeconomic terms than the general rebate approach. The price and allocation approach is the most inefficient in that allocations are based on historical usage or queuing, both of which impose enormous economic and social costs.

> *"The coupon rationing and tax-rebate approaches allow efficient allocation of gasoline, assuming the government does its job perfectly."*

None of the approaches mitigate the supply side macroeconomic costs -- higher prices and recession -- associated with declining world oil production. The true price of oil, whether measured in terms of queues, coupons, or dollars, will be higher. Emergency allocation approaches and monetary and fiscal policies can, however, affect the demand side macroeconomic costs. The sudden, massive movement of funds into the oil market could sharply reduce output in non-oil sectors of the economy. This oil price drag is associated most dramatically with the gasoline tax and rebate and the general rebate approaches. Further, if an allocation approach can somehow exclude oil price increases from the CPI, labor agreements, contracts, and government entitlement programs that are indexed to the CPI will not escalate as rapidly as if oil price increases were included. Price controls keep oil price increases out of the CPI by requiring payment in terms of time spent waiting in gasoline lines, while rationing keeps gasoline price increases out of the CPI by creating a second currency.

While unquestionably an important criterion, equity is to a large degree a matter of perception. None of the four approaches will be perceived as fair to all groups. The general rebate approach may be perceived as the least equitable, since it allows prices to rise and enables oil companies to charge what the market will bear. Wealth will be transferred from oil consumers to oil producers, both domestic and foreign. Even if a windfall profits tax is enacted to capture most of the windfall, and even if all revenues are rebated to consumers, the public's perception will likely be one of oil companies making money at the expense of consumers. To a lesser degree, the gasoline tax and rebate approach will likely be perceived as unfair because it involves an explicit tax on consumer products. Paradoxically, the gasoline rationing approach, which is as fair as the tax and rebate approach, may

> *"Equity is to a large degree a matter of perception. None of the four approaches will be perceived as fair to all groups."*

be perceived as the most equitable means of allocating gasoline supplies. Even though the two approaches would probably lead to similar distributions of available gasoline supplies, the possession of a coupon confers a "right" to a gallon of gasoline in a way that currency does not. Even the coupon gasoline rationing approach will be perceived by some -- those who do not own automobiles, for example -- to be unfair. The arbitrary nature of the first-come, first-served gasoline lines resulting from this approach cannot be perceived as fair for any extended period of time.

The nature and magnitude of practical problems associated with each approach are important considerations. Three of the approaches -- price controls, rationing, and gasoline tax and rebate -- would require government to implement oil price controls. The imposition of these controls with the attendant entitlements program simultaneously with the imposition of gasoline rationing could strain some government resources. Regarding time to implement, the ability of the general rebate approach to allocate oil and oil product supplies quickly appears to give it significant advantages over other options.

The four approaches require different amounts of information on which to base decisions. The rationing approach requires projections of the volumes of gasoline available several months in advance. The gasoline tax and rebate approach requires estimates of the size of the tax necessary to equate demand with supply; such information is not now available and is not likely to be reliable even if collected. The general rebate approach requires relatively less data for the decisions required. Finally, the ease of dismantling an allocation system following a disruption must be considered. Approaches requiring any form of price controls may prove more difficult to phase out than ones that do not.

Finally, it should be abundantly clear from the foregoing that rationing is easy to say and extraordinarily difficult to implement. There is no rationing approach that stands out as simple and fair. [67] In all cases, the required government organization to implement and manage any of the approaches will be very significant. Government bureaucracies will grow and politics will surely create special preferences.

C. Carpooling

Carpooling represents a significant and rapid means of reducing the

"Three of the approaches -- price controls, rationing, and gasoline tax and rebate -- would require government to implement oil price controls."

> **"Rationing is easy to say and extraordinarily difficult to implement."**

consumption of gasoline and other petroleum products. The shared use of vehicles by multiple occupants applies not only to people commuting to and from work but also to a range of other people-moving activities, such as shopping, athletic events, religious events, social events, etc.

Considerable progress has been made in supporting carpooling over the last 35 years, but it developed more from the discomfort of traffic congestion than from the desire to lower gasoline consumption. Businesses have given preferential treatment to employees who car-pool by providing subsidies, operating company van-pools, providing free parking, etc. The origin of company subsidies has sometimes been government tax treatment or programs to relieve traffic congestion. One effective government program that encouraged car-pooling is the designation of roadways as high-occupancy vehicle (HOV) lanes, reserved during certain hours for vehicles with multiple occupants.

According to the U.S. Department of Transportation, [68] the U.S. recently averaged 1.63 occupants per vehicle/mile, which varied by type of vehicle:

- Cars - 1.58
- Vans - 2.20
- SUVs - 1.74
- Pickup trucks - 1.46
- Other trucks – 1.20
- Motorcycles – 1.27

Average occupancy varied by the purpose of the trip:

- Home to work – 1.14
- Work-related – 1.22
- Shopping, family, and personal – 1.81
- Church, school – 1.76
- Social, recreational – 2.05
- Other – 2.02

> **"Carpooling represents a significant and rapid means of reducing the consumption of gasoline and other petroleum products."**

"Clearly, the most wasteful practice is the single-driver trip to and from work."

Clearly, the most wasteful practice is the single-driver trip to and from work. Consider a what-if situation. In the U.S. in 2007, light-duty transportation vehicles accounted for about 9 million barrels per day of petroleum consumption. Twenty-seven percent of the miles driven were to and from work. Therefore, around 2.4 million barrels a day were used to get to and from work. During those trips there was a weighted occupancy rate of 1.14 persons per vehicle mile. If the occupancy rate was increased 50% to 1.71 persons per vehicle mile, the total miles driven would be decreased by 50%, resulting in a decrease in petroleum consumption of up to roughly a million barrels per day of oil consumption.

While mass transportation is growing in some urban areas in the U.S., it only accounts for about five percent of trips to work. Private vehicle trips to work in a carpool account for roughly 11 percent of trips, while single-driver trips account for three-quarters of all trips. If carpooling were expanded, significant reductions in gasoline demand would result. Indirect benefits would accrue also, including a decrease in traffic congestion, a decrease in vehicle emissions, and more money in people's pockets for discretionary spending.

Significant increases in average vehicle occupancy across the spectrum of activities in the U.S. alone have been estimated to have the potential of around a million barrels per day of savings. Elsewhere in the world, significant savings are also possible, but the savings would be strongly influenced by the existing use of public transportation and variations in population densities.

In addition to individuals voluntarily changing their driving habits, organizations can stimulate carpooling by the following:

- Providing flexible work hours that allow easier work matching among employees
- Subsidizing carpool vehicles and pool drivers
- Providing preferential parking for carpools
- Providing or supporting rideshare matching services to help people find convenient carpools

Governments could:

- Offer direct financial incentives for car pools
- Provide greater tax savings for people who carpool

> ## "Government mandating of carpooling would not be an easy task."

- Increase the availability and use of designated HOV lanes
- Create and encourage ad hoc carpool pick-up and drop-off areas
- Mandate trip reduction programs at large companies
- Provide rideshare matching programs

Government mandating of carpooling would not be an easy task, because of the following:

- Not everyone is located in places where carpooling is practical, e.g., rural areas, so a simple mandate would be inherently unfair to many.
- Employment patterns will change as oil-decline-induced recession causes shifts in business activities and commuting patterns.
- Monitoring and enforcing mandated carpooling will be both expensive and time consuming for law enforcement.

In conclusion, carpooling is a ready means for reducing oil product consumption. The high cost of fuels and fuel shortages will provide a natural impetus for people to carpool without government mandate. Whether governments decide to intercede, and to what degree, is not is not now knowable.

D. Telecommuting

Telecommuting is a work arrangement where employees work from home or some other work site, physically separated from the normal work place for part of the workday or part of the work week. Telecommuting is feasible where an employee's work activities can be performed remotely and is facilitated by computers, the Internet, telephones, and remote interactions via video links. By reducing or eliminating the need to travel, gasoline consumption can be reduced.

Telecommuting is a growing trend in many parts of the world. When world oil production goes into decline, oil product prices will escalate and shortages will ensue, so that telecommuting will become even more attractive.

Estimates are that in the U.S., 1) 2.5 million employees consider home their primary place of business, 2) as many as 8 million Americans (almost 6 percent of the workforce) telecommute to jobs from their homes on either

> ## "Telecommuting is a growing trend in many parts of the world."

a part-time or full-time basis, and 3) more than 17 million occasionally work remotely. [69] There is no doubt that these numbers could be increased significantly.

It is obvious that there are many types of work where employees cannot telecommute because they must interact directly with others. These include retail sales, hospitals, emergency services, police and fire fighting, certain kinds of customer services, merchandise pickup and delivery, refuse pickup, etc. In many organizations, career advancement is dependent on working directly with other people in common locations. In these situations telecommuters could be at a significant disadvantage. In addition, telecommuting requires surveillance by management to ensure that telecommuters are performing their jobs in the manner that their employers expect.

Beside gasoline savings, some of the other benefits of telecommuting include the following:

- Businesses require less office space (structures and real estate), electricity, maintenance, parking, etc.
- Employees save on transportation, food, clothing, potential daycare costs, etc. and decrease their potential for transportation-related accidents.
- Employees have more time with their families and for recreation.
- Reduced commuter traffic decreases the cost of traffic congestion and the cost of road construction and maintenance, and reduces the amount of vehicle emissions.

In the U.S., telecommuting gained substantial government recognition in 1996, when the Clean Air Act required companies with over 100 employees to encourage carpools, telecommuting, and other practices in order to reduce air emissions. The U.S. Federal Government encourages telecommuting as follows:

- In 2004 Congress encouraged Federal Agencies to provide telecommuting options to eligible employees and threatened to withhold agency funding where successful programs were not put in place.
- The U.S. General Services Administration had over 10 percent of its workforce several years ago telecommuting and now has a goal to reach 50 percent.
- The National Science Foundation announced that approximately one-third of its employees participated in telework regularly. [70]
- Telecommuting is seen as a decentralization of government activities, which offers other national security-related benefits.

> **"Around the world significant telecommuting opportunities exist."**

Around the world significant telecommuting opportunities exist. Organizations can encourage greater participation on their own and governments working with employers can develop additional opportunities. The degree to which governments institute mandates is an open question.

One caution: The viability of telecommuting depends on maintaining the integrity of the electric power system, the phone system, and the Internet. Severe damage to any of those systems by extreme weather or terrorism would do significant economic harm, so related vulnerabilities would need to be monitored and minimized.

XI. The Best That Physical Mitigation Can Provide

A. Background

1. Introduction

In a 2005 study for the U.S. Department of Energy, the three of us considered the looming world oil shortage problem and developed scenarios for the most that a worldwide crash program on physical mitigation might accomplish. [71] Why? Because a worldwide crash program of physical mitigation represents the very best that is likely to be physically possible. Interestingly, our study has been presented worldwide numerous times over the ensuing years and has endured as a major reference with analysts and the media.

In our 2005 study, we considered three time-dependent scenarios for worldwide crash program mitigation: 1) An effort initiated 20 years before the onset of world oil production decline; 2) An effort started 10 years ahead; and 3) An effort initiated when the decline becomes obvious. When our study was conducted, we did not have a sound basis for judging when the onset of world oil production decline might occur, so we investigated the three scenarios to elucidate an array of possibilities.

> *"A worldwide crash program of physical mitigation represents the very best that is likely to be physically possible."*

Today, we believe that world oil production will likely go into decline within a matter of years, so consideration of efforts initiated 10 or 20 years beforehand is of little practical value. In addition, because the onset of the decline of world oil production ("peak oil") is still a controversial subject and not yet taken seriously by most governments, it is clear that significant physical mitigation is unlikely to begin before the problem is obvious.

In our 2005 study, we assumed that world oil production might follow the sharply peaked pattern that occurred in the U.S. Lower 48 states. The idea that world oil production might reach a multi-year plateau and then go into decline was not yet a mainline scenario. Today, the situation is different; world oil production has been on a fluctuating plateau since mid 2004, and a

> *"The actual decline rate is a critical parameter in the mitigation problem, because the higher the rate, the more serious the problem and the longer it will take to overtake a decline that has a head start on physical mitigation."*

decline from the current plateau is now our most likely scenario.

Previously, we assumed that the decline rate for world oil production would be 2% per year over a very long period. That was the decline rate for oil production in the U.S. Lower 48 states, and it seemed that a 2% decline rate was a reasonable assumption for world oil production. Today, credible analysts postulate much higher decline rates. Indeed, the actual decline rate is a critical parameter in the mitigation problem, because the higher the rate, the more serious the problem and the longer it will take to overtake a decline that has a head start on physical mitigation.

2. World Oil Production Decline Rates

There are a number of clues as to the likely decline rate of world oil production. First, there are the rates that have occurred in very large oil producing regions that are well past maximum production. Second, there are rates estimated by various analysts. Third, there is the impact of exporter country policies, which could result in decline rates higher than what is physically possible (exporter withholding). Some decline rate estimates are summarized in Table X-1.

Table XI-1 Approximate Decline Rates in Large Regions

Actual U.S. Lower 48 States after 1970............................... 2%
Actual Europe after its plateau production.........................5-6%
Estimate from Campbell...2%
Estimate from Skrebowski..2.5%
Estimates from Robilius's four cases..........2%, 3%, 3.5%, & 4.5%
Estimate from the National Petroleum Council...................4-7%

Overall range...............2-7%

While the upper limit on decline rate will be geology, human factors can and often do overshadow what nature allows. For example, faster decline rates in various countries can result from bad oil field management, inadequate investments, and poor or selfish government decision-making.

Oil field investments emerged as an issue in 2006, when the IEA Executive Director discussed the subject: [72]

"(The IEA) WEO 2006 identifies under-investment in new energy supply as a real risk", said Claude Mandil, IEA Executive Director. To quench the world's thirst for energy, the IEA Reference Scenario projections called for a cumulative investment in energy-supply infrastructure of over $20 trillion in real terms over 2005-2030 - substantially more than was previously estimated. It is far from certain that all this investment will actually occur. There has been an apparent surge in oil and gas investment in recent years, but it is, to a large extent, illusory. Drilling, material and personnel costs in the industry have soared, so that in real terms investment in 2005 was barely higher than that in 2000.

"In interviews associated with the WEO 2006 release, IEA executives further affirmed the impending difficulties: 'This energy future is not only unsustainable, it is doomed to failure,' because of, "underinvestment in basic energy infrastructure,' '...In short, we are on course for an energy system that will evolve from crisis to crisis.' "

So it is possible -- some would say probable -- that geologically dictated decline rates will be steepened by under-investment. But as oil prices increase in the ensuing years, the impact of past under-investment might moderate somewhat, if exporters were to accelerate oil field investments, which they may not.

If world oil production were controlled by the International Oil Companies (IOCs), such as ExxonMobil, Chevron, Shell, ConocoPhillips, etc., investments would almost certainly be forthcoming, driven by the profit motive. But the IOCs are not in control; the majority of world oil reserves and production are in the hands of National Oil companies (NOCs), such as Saudi Aramco, the Iranian National Oil Company, Kuwait National Oil Co., Petróleos de Venezuela, etc.

> *"Government-controlled companies actually control the majority of both current production (more than 52% in 2007) and proven reserves (88% in 2007)."*

According to EIA: [73]

"Although investor-owned oil companies are often thought of as those most responsible for world oil production, (foreign) government-controlled companies actually control the majority of both current production (more than

52% in 2007) and proven reserves (88% in 2007), one indicator of future production potential.

"National oil companies that operate as an extension of the government or a government agency, including Saudi Aramco (Saudi Arabia), Pemex (Mexico), and PDVSA (Venezuela), support their government's programs either financially or strategically. ... These companies do not always have the incentive, means, or intention to develop their proven reserves at the same pace as commercial companies. Due to the diverse situations and objectives of the governments of their countries, these national oil companies pursue a wide variety of objectives that are not necessarily market-oriented, such as employing their citizens, furthering a government's domestic or foreign policy objectives, generating long-term revenue, and supplying inexpensive domestic energy."

For those who might want to blame Exxon and other major oil companies for future oil shortages, the new reality is bad news; as big as Exxon is, it is a small player on the world oil stage, responsible for only about 3% of total world oil production, so other entities will have to be found to blame.

As an example of what governments that control national oil companies might do, consider the following. The massive increases in world oil prices from the early 2000s to the price peak in 2008 provided oil-exporting countries with enormous financial windfalls. In the future, when it becomes clear that total world oil production is in decline -- "post-peak" -- oil prices will once again escalate dramatically, providing vast new revenues to oil exporters. Many exporting countries will have more than enough money to meet their needs at that point. In addition, it is likely that a number of oil exporters will realize the frailty of their own longer-term resource circumstances and decide to hold back oil production to stretch their oil endowment for future generations. We call that possibility the Oil Exporter Withholding Scenario. [74]

Oil withholding could easily be done subtly, so as to not inflame oil importers. There is relatively little that shortage-prone oil importers will be able to do to alter exporter withholding, because exporters are sovereign nations with the right to manage their resources as they see fit. One might ask about the use of the military to spur countries to increase their exports. Military intervention is not likely a viable option, based on a senior level study a few years ago that concluded, "Military options offer little recourse in the event of a supply crisis." [75]

It is axiomatic that throttling back oil production can be done almost instantly. Many countries have done so on occasion; it is one tool that OPEC has used in their efforts to manage prices over the years. On the other hand, expanding production on a significant scale requires up to a decade, assuming the oil reserves actually exist for development.

Where does all of this leave us? We believe that two cases bracket future world oil production decline:

I. **BEST CASE:** A number of years of relatively flat world oil production (the existing plateau) before the onset of a decline rate of 2% per year.

II. **WORST CASE:** A few years of relatively flat world oil production (the existing plateau) followed by the onset of a decline rate of 4% per year.

Remember that world oil production decline rates will increase over time as more and more oil fields go into decline. In our rough estimates, we ignore that fact for simplicity, but growing world oil production decline rates are nevertheless a fact of life.

3. Elements of a World Oil Production Decline Analysis

A major challenge in performing a meaningful analysis of the best-case mitigation of world oil production decline is the selection of the most important variables, while making reasonable assumptions on other important factors. In the following, we adopt the time-tested approach used in our 2005 DOE study. An exact analysis is impossible, because of the multitude of complexities and unknowns, but it is possible to develop rough estimates, and in the process, learn a great deal about the challenges ahead.

We recognize that some readers may find the use of approximations and estimates unsatisfying. We can only say that over the decades of our experience, rough estimate calculations have frequently proven robust in approximating complex situations in engineering and economics.

To repeat, our interest in a worldwide crash program is based on the belief that it represents the upper limit on what can be physically accomplished. We assume an ideal crash program -- one that starts instantaneously, going from business-as-usual one day to full-blown implementation the next day. Such an overnight scale-up is impossible for reasons that we will discuss later, but the step-change assumption is useful in our efforts to understand the best that is possible.

4. The Delayed Wedge Approximation

In our 2005 study, we used a "delayed wedge" method to approximate the scale and growth of each mitigation option, because the method captured key elements of how things work, and it simplified our analysis. Wedges are composed of two parts. The first is the startup phase that occurs before the appearance of tangible results. For example, in the case of the construction of Coal-To-Liquids (CTL) plants, a period of time is required to design, plan, and build the plants before they can begin to produce useful liquid fuels. In the case of the manufacture of very fuel-efficient cars, assembly plants must be reconfigured and suppliers must manufacture and deliver vehicle components before the first vehicles can be assembled and made available for sale.

After the startup phase, wedges are assumed to grow linearly with time to reflect the liquid fuel savings or liquid fuels production that accumulates. The delayed wedge pattern is shown in Figure XI-1, where the horizontal axis is time and the vertical axis is market penetration, measured in millions of barrels per day of savings, as in the case of efficient vehicles, or liquid fuels production, as in the case of CTL plants. We considered a 20-year period because major impacts require a long time, as we will demonstrate. We recognize that the further out in time that we go with our approximations, the less accurate they will be. While these limitations may seem troublesome to those not used to developing rough estimates, remember that our interest is in identifying major trends, which this formalism facilitates. In fact, to forecast in greater detail is impossible, because of the myriad of unknowns and complexities.

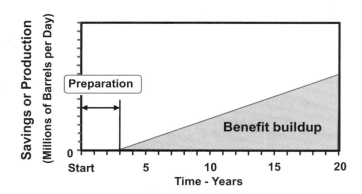

Figure XI-1. Delayed wedge approximation for various mitigation options

5. Selection of Major Physical Mitigation Options

The criteria we used for our selection of the most viable physical mitigation options were as follows:

1. The option must be capable of contributions on a massive scale -- millions to tens of millions of barrels per day worldwide within a decade or two.

2. The option must involve technology that is now commercial or near commercial. Technologies still in the research and development stage do not qualify until they are demonstrated at large scale.

3. The option must either save or produce liquid fuels suitable for widespread use in existing capital stock, e.g. cars, trucks, buses, airplanes, ships, industries, etc

4. Fuel production technologies must be liquid-fuel efficient, which means that the volumes of liquid fuels produced must be many times greater than the liquid fuels consumed in construction and operation of related facilities. Producing just a little bit more liquid fuels than are consumed in a process is a waste of time and effort. [76]

5. The option must be environmentally reasonable but not necessarily ultraclean. Every energy option has shortcomings. Compromises are essential, because no technology is perfect! As is often said, "The best is the enemy of the good."

6. The feedstock resources (heavy oil, coal, natural gas, etc.) must be plentiful somewhere in the world. If not, millions to tens of millions of barrels of liquid fuels will not be possible.

7. Technologies that produce or save electricity in the near-term are not of interest for solving a liquid fuels shortage problem. However, electricity production options can be extremely valuable in their own right and could be of great value in the long term, when greater electrification of the transportation sector can be developed.

6. The Options

For the purposes of our rough estimates of what worldwide crash program mitigation might achieve, we considered the following:

- Fuel efficient transportation,
- Heavy Oil/Oil sands,
- Coal-to Liquids (CTL),
- Enhanced Oil Recovery (EOR) and
- Gas-to-Liquids (GTL)

Before discussing each in some detail, here are some options that we chose to ignore, along with our reasoning:

1. **Shale Oil:** While the world has a huge resource of shale oil that could be processed into substitute liquid fuels, related production technologies are not yet commercial and not ready for deployment.

2. **Biomass-to-Liquids**: Ethanol from corn is currently utilized in the U.S. transportation market, because of government mandates and subsidies. Corn-ethanol is not commercially viable on its own and would disappear if it had to fairly compete with other options. Furthermore, it may not even be energy positive; the technical literature in this case is ambiguous. Biodiesel and cellulosic conversion are not yet commercially viable. Continuing research and development may change the biomass outlook, but so far, no biomass options meet our criteria.

3. **Fuel Switching in the Electric Power Sector:** As a result of the oil shocks of the 1970s, the U.S. and other nations switched most of their oil-fired electric power production to other fuels. However, the switch was not complete, so opportunities for switching electric power generation away from liquid fuels certainly exist. For significant world scale impact, alternate fueled large electric power generation facilities would have to be retrofitted or constructed from scratch on a massive scale to provide makeup for the loss of the oil-fired electric power production. We know of no way of estimating how fast such conversions might happen, because of the different economic circumstances in each country where opportunities exist. However, for world-scale oil-market impact, related facility construction would require the kind of decade-scale time periods required for oil decline mitigation.

4. **Nuclear Power, Wind and Photovoltaics:** These technologies produce electric power, which is not a near-term substitute in existing transportation equipment that is built to operate on liquid fuels. In the future, it is likely that a massive shift away from liquid fuels to electricity will occur in a number of applications.

5. **Electrified Railways:** The majority of railway power in the U.S. is provided by diesel fuel, which opens up the potential for electrification. However, such an undertaking would require the construction of new electric power plants, transmission lines, and electric locomotives. Since existing diesel locomotives use electric drive, some retrofits are conceivable. However, because diesel fuel use in trains is only about 0.3 million barrels per day in the U.S., electrification of U.S trains would not have a major impact on

total U.S. liquid fuel consumption. Elsewhere in the world, railway electrification is already widespread, so the overall world impact would likely be modest, which is not to deny its utility.

6. **Building Heating:** Worldwide, a significant number of homes and commercial buildings are warmed by heating oil / or Liquefied Petroleum Gas (LPG). In principle these facilities could be switched to natural gas or electric heating, freeing up liquid fuels for transportation. Analysis of this path is complicated and dependent on a large number of personal and national circumstances. Keep in mind that switching on a large scale would require the construction of compensating natural gas and/or electric power source facilities and delivery infrastructure, which cannot happen quickly on the scale of world oil physical mitigation. Directionally, these conversions could be useful liquid fuels savers. However, per-facility costs are not trivial and are up to the property owner, who is often unable to afford the conversion.

7. **Miscellaneous Activities**: Many smaller scale efficiencies and liquids production options will happen over the next few decades. We could have included a wedge for "all other," but it would have been strictly speculative. Furthermore, no matter what the new technologies, the magnitude of their contributions is certain to be much less than the technologies that we considered, because of the inherently huge scale of required physical mitigation.

B. More Efficient Vehicles as a Mitigation Pathway

1. Cars and Trucks

a. Introduction

Automobiles and light trucks – often called Light Duty Vehicles (LDVs) -- are large consumers of gasoline worldwide. Emerging technologies are now, and soon will be, available to provide much higher mileage, thereby saving gasoline. These include Hybrid Electric Vehicles (HEVs), Plug-in Hybrid Electric Vehicles (PHEVs), and Electric Vehicles (EVs).

b. Hybrid Electric Vehicles

Hybrid Electric Vehicles (HEVs) are commercial today. They involve a small gasoline engine and a large battery pack to provide much higher fuel economy than a larger sized gasoline engine can by itself. During acceleration, both the engine and the battery typically provide power to the wheels. The battery is charged by the gasoline engine during non-acceleration periods and by energy captured as the vehicle slows down and when the brakes are applied.

The most popular HEV has been the Toyota Prius. Other HEVs with growing sales include the Honda Insight, the Honda Civic Hybrid, the Ford Escape Hybrid, the Ford Fusion Hybrid, Toyota Camry Hybrid, and a number of Lexus offerings. The number of HEVs on the road worldwide in 2008 was near 450,000 of which 62% were in the U.S. [77] Various manufacturers plan to introduce additional HEVs in the near future.

HEVs are at least several thousand dollars more expensive than comparable gasoline-powered vehicles. While HEV sales increase during periods when gasoline prices were high they have been tiny compared to the entire fleet size. Following the onset of the decline in world oil production, interest in HEVs is certain to increase, but the ensuing recession will hurt sales, as recently demonstrated.

> *"Costs of plug-in hybrid electric cars are high -- largely due to their lithium-ion batteries -- and unlikely to drastically decrease in the near future."*

c. Plug-In Hybrid Electric Vehicles

An evolving option for even more fuel-efficient automobiles is the Plug-in Hybrid Electric Vehicle (PHEV). These automobiles are like HEVs but have larger battery packs. PHEV batteries are charged by plugging the vehicle into a source of electric power, envisioned to often occur at home at night.

One of us (RLH) was a member of a National Academies committee that recently evaluated the viability of PHEVs and the outlook for them to impact future liquid fuels consumption. The following is from a summary of the findings of the committee: [78]

> "Costs of plug-in hybrid electric cars are high -- largely due to their lithium-ion batteries -- and unlikely to drastically decrease in the near future ... Although a mile driven on electricity is cheaper than one driven on gasoline, it will likely take several decades before the high upfront (vehicle) costs decline enough to be offset by lifetime fuel savings. Subsidies in the tens to hundreds of billions of dollars over that period will be needed if plug-ins are to achieve rapid penetration of

the U.S. automotive market. Even with these efforts, plug-in hybrid electric vehicles are not expected to significantly impact oil consumption or carbon emissions before 2030.

"The report considers plug-in hybrid electric vehicles that can operate on electricity for 10 or 40 miles. The PHEV-10 is similar to the Toyota Prius but with a larger battery. The PHEV-40 is similar to the Chevrolet Volt...The lithium-ion battery technology used to run these vehicles is the key determinant of their cost and range on electric power. Battery technology has been developing rapidly but steep declines in cost do not appear likely over the next couple of decades, because lithium-ion batteries are already produced in very large quantities for cell phones and laptop computers. In the first generation of production, the PHEV-10 battery pack is estimated to cost about $3300, and the PHEV-40 battery pack about $14,000. While these costs will come down, a fundamental breakthrough in battery technology, unforeseen at present, would be needed to make plug-ins widely affordable in the near future.

"According to the committee that wrote the report, the maximum number of plug-in electric vehicles that could be on the road by 2030 is 40 million, assuming rapid technological progress in the field, increased government support, and consumer acceptance of these vehicles. However, factors such as high cost, limited availability of places to plug in, and market competition suggest that 13 million is a more realistic number, the report says. Even this more modest estimate assumes that current levels of government support will continue for several decades.

"Most of the electricity used to power these cars will be supplied from the nation's power grid. If powered at night when the demand for electricity is lowest, the grid would be able to handle the additional demand for millions of plug-in hybrid electric vehicles, the report says. However, if drivers charge their vehicles at times of high demand, ... the additional load could be difficult to meet unless new capacity is added. Smart meters, which bill customers based on time of use, may be necessary in order to encourage nighttime charging. In addition, some homes would require electrical system upgrades to charge their vehicle, which could cost over $1,000.

"Relative to hybrid vehicles, plug-in hybrid electric

vehicles will have little impact on U.S. oil consumption before 2030, especially if fuel economy for conventional vehicles and hybrids continues to increase past 2020. PHEV-10s save only about 20 percent of the gasoline an equivalent hybrid vehicle would use, the report says. If 40 million PHEV-10s are operating in 2030, they would save about 0.2 million barrels of oil per day relative to less expensive hybrids, approximately 2 percent of current U.S. daily light-duty vehicle oil consumption. More substantial savings could be seen by 2050. PHEV-40s, which consume 55 percent less gasoline than hybrids, could have a greater impact on oil consumption."

"PHEVs will be considerably more expensive than HEVs and will almost certainly require large, long-term government subsidies."

Thus, PHEVs will be considerably more expensive than HEVs and will almost certainly require large, long-term government subsidies, if they are to have a major impact on future liquid fuels consumption. Two caveats to the Academies conclusions are important:

1) A major Academies study assumption was that the economic environment in the U.S will be much the same in the long-term future as it was prior to 2008. If the onset of the decline in world oil production happens within a matter of years, as we expect, then business-as-usual will be interrupted in a major way, and there will be an annually deepening recession until physical mitigation takes hold. In such a situation, automobile purchases, particularly for much more expensive vehicles like PHEVs, will almost certainly be diminished, as demonstrated by the drop in car sales during the recent recession. We believe this will be the case even if a new "cash-for-clunkers" program is instituted, because the cost of a program that could significantly impact car purchases on the scale that have a large national impact is likely beyond what governments will be able to afford.

2) Future oil prices assumed in the Academies study were based on EIA forecasts, which assume business-as-usual and do not foresee the decline in world oil production. In our scenario for the future, the oil production decline will result in

oil prices (and thus gasoline prices) that are much greater than in the EIA forecasts, which might mean that lower government subsidies may be needed in a world where gasoline is very expensive and may often be in shortage.

In conclusion, the attractiveness of PHEVs after the onset of the decline in world oil production will be buoyed by high fuel costs, while sales will be dampened by annually increasing recessionary pressures. Forecasting how those factors will balance out over time is nearly impossible in our opinion; there are just too many unknowns and variables.

d. Electric Vehicles

The EV1 electric automobile was produced by General Motors and leased in California during the period 1996-1999. GM was able to lease only 800 vehicles and had reported production costs of roughly a billion dollars over four years. The program was not a commercial success, and in 2002, all EV1s were repossessed and the program was discontinued.

With the emergence of the lithium-ion battery, interest in electric vehicles (EVs) has been rekindled, and a number of manufacturers have announced plans for EV production. In the U.S., Tesla Motors sells an expensive roadster with a claimed driving range over 240 miles between charges. However, at a cost of over $100,000, its market potential was severely limited. Subsequently, Tesla announced plans to introduce a lower priced four-door sedan.

Other manufacturers have a number of electric vehicles in the prototype stage, and some have promised commercial production within a few years. For example, Nissan has heralded its LEAF all-electric, four-door sedan. At the time we wrote this book, Nissan's web site indicated that, "exact specs of the LEAF are still under development." As of December 2009, Nissan also stated that, "we're unable to give an exact price, but we're targeting a price in the range of other typical family sedans."

We expect a number of manufacturers to market electric vehicles in the near future. It is important to recognize that EVs, unlike PHEVs, have no backup power, so when the battery charge is depleted, owners will have to have their EVs either towed to a charging station or receive a quick charge from special charging trucks, yet to be deployed.

It is likely that EVs will find a following with people who drive much less than the vehicle range each day. EVs are likely to be small, thereby requiring less expensive batteries. Otherwise they are certain to be more expensive

> **"It is important to recognize that EVs, unlike PHEVs, have no backup power."**

than comparable standard gasoline fueled vehicles, hybrids, or PHEVs. With little commercial EV experience, the outlook for widespread EV deployment is very difficult to estimate.

e. Fuel Efficiency Mandates

Higher government-mandated vehicle fuel efficiency requirements are certain to be an element in the mitigation of world oil shortages. The result may be the more rapid deployment of smaller and more efficient vehicles. However, the market penetration of these technologies cannot begin quickly, because of the time and effort required for manufacturers and suppliers to retool their factories for large-scale production and because of the slow turnover of existing light duty vehicles.

The imposition of stricter fuel efficiency standards have been demonstrated to be effective, based on the U.S. experience following the 1975 imposition of Corporate Average Fuel Efficiency (CAFE) requirements. Recently, the U.S. government imposed new, more restrictive fuel economy requirements, and so the U.S. has made a significant step along the path of more fuel-efficient light duty vehicles (LDVs). While the recent U.S. action is positive, we believe that the pace of implementation demanded by the recent CAFE program is still not adequate for the demands of declining world oil production.

For our current estimates, we assume that much higher CAFE standards will be implemented in many countries around the world. On the other hand, the recent recession clearly demonstrated that severe economic downturns can dramatically reduce the number of new vehicles that the public is able to purchase. After reviewing our earlier study, we believe that the mileage estimates in our 2005 DOE study were somewhat less aggressive than we now judge to be possible. On the other hand, we believe that the greater mileage provided by recent technology advances will be muted by lower sales, balancing out close to our earlier assumptions.

> **"The imposition of stricter fuel efficiency standards has been demonstrated to be effective, based on the U.S. experience following the 1975 imposition of Corporate Average Fuel Efficiency (CAFE) requirements."**

The U.S. currently has about 25 percent of total world vehicle registrations, but consumes nearly 40 percent of the liquid fuels used in transportation worldwide. However, China recently overtook the U.S. in new car sales and will undoubtedly continue to expand.

We could not find credible forecasts of the potential impacts of increased worldwide vehicle fuel efficiency standards, so we assumed that the impact of enhanced vehicle fuel efficiency standards in the rest of the world will be about equal to that in the U.S. In total, the worldwide impact of increased vehicle fuel efficiency standards might thus yield a savings of 1 million barrels per day of liquid fuels 10 years after legislation is enacted; 3 million barrels per day 15 years after legislation is enacted; and 6 million barrels per day 20 years after legislation is enacted.

It would be preferable to have a more detailed study upon which to base our estimates, however, we believe our assumptions are reasonable for the purposes of the current analysis. Remember, our goal is to roughly estimate what the character and the sum total of all saving and production options might be, which means that modest variations in estimates of the contribution of each option are less important than the final result, which will necessarily be a rough estimate.

Retooling the automobile industry to massively manufacture high mileage vehicles is assumed to require three years from the decision to go ahead, because of the time needed to modify production plants and for suppliers to make the changes needed to produce vehicle components. That estimate is in line with experience in the industry, but it is possible that the time could be reduced somewhat.

Our estimate for the contributions to world fuel saving from more restrictive fuel economy standards is shown in Figure XI-2.

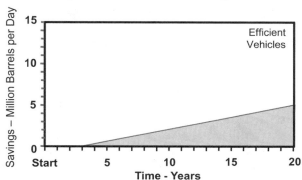

Figure XI-2. Worldwide fuel savings from much higher fuel efficient, light duty vehicles. The full wave of new vehicles is assumed to enter the market in three years after a decision to massively retool.

When we began our 2005 study, we expected a much higher contribution from higher efficiency vehicles, but our analysis led us to a more modest result. We reiterate that the experience of the recent recession strongly points to greatly reduced vehicle sales during an ever-deepening oil shortage-driven recession.

f. Efficiencies in Other Vehicles

On the distant horizon, innovations in aircraft design may result in large fuel economy improvements. Boeing claims that its new 787 aircraft uses 20% less fuel than any other airplane of its size. A recent National Academies study suggested that "jet engine and aircraft technology has the potential to improve the efficiency of new aircraft by up to 35 percent over the next two decades," [79] but such changes cannot happen quickly due to the need to ensure that resulting aircraft are safe and durable. Again, during a deepening recession, the rate of new aircraft purchases is certain to be reduced, so overall impacts are very difficult to forecast.

In the same report the Academies commented on improved truck efficiencies as follows: "Reductions of 10-20 percent in the fuel economy of heavy- and medium-duty vehicles appear feasible over a decade or so. A broad examination is needed of the potential for improving the effectiveness of the freight system to reduce energy consumption further." We do not doubt the potential, but the timing and penetration of improved technologies is very much open to question.

On the basis of these considerations, we expect important improvements in other transportation vehicles, but not as large as what we believe is possible in Light Duty Vehicles.

C. Heavy Oil & Oil Sands as a Mitigation Pathway

Because of their high viscosity, many oils do not readily flow out of underground reservoirs. Such oils are called "heavy," and they usually contain significant levels of impurities, which must be removed by extensive refining. Still, there are large quantities of these oils in the world, and while they are more expensive to produce and refine than most light oils, they can be economic at oil prices in the range of $40-80 per barrel. These unconventional oils are variously called heavy oil, bitumen, oil sands, and tar sands. We believe heavy oils will play a growing role in satisfying the world's needs in the future, because they are plentiful and have been proven commercially.

The largest heavy oil deposits exist in Canada and Venezuela, with smaller quantities in Russia, Europe, and the U.S. Production of heavy oils

> **"We believe heavy oils will play a growing role in satisfying the world's needs in the future, because they are plentiful and have been proven commercially."**

has been ongoing for decades, so there is no question regarding commercial viability. While there are issues related to their environmental impacts, these issues can and will be managed in an oil-starved world, because oil shortages and very high oil prices can impact priorities.

Canadian production of its heavy oils, called oil sands or tar sands, was roughly 1.3 million barrels per day in 2009. A crash program expansion of their already aggressive production plans faces a number of challenges, including the need for massive supplies of auxiliary energy, e.g., natural gas, huge land and water requirements, careful environmental management, and the harsh wintertime climate in the region.

Our assumptions for a crash program of Canadian oil sands are as follows:

1. Accelerated production might begin three years after a decision to proceed with a crash program, based on the fact that the country already has significant production underway.

2. The recent Canadian vision was to produce roughly 4 million barrels per day from oil sands by 2020. That amounts to an additional 2.7 million barrels per day in roughly 10 years or an additional average of 0.27 million barrels per day per year. Our crash program expansion of their already aggressive plan assumes an acceleration by a factor of two. Thus, after a 3-year startup period, the crash program increment would yield an additional 2.7 million barrels per day by year 13 and about 5 million barrels per day after 20 years. The resultant Canadian production would total well above 5 million barrels per day. Such a high level may not be possible, if a Canadian oil sands crash program analysis from Uppsala University is correct. [80] Time and more detailed analysis will tell.

Venezuela has very large heavy oil reserves, and the government claimed production of roughly 0.6 million barrels per day in 2009. That number must be viewed with skepticism, because various observers believe the actual level to be lower. The political situation in Venezuela is currently chaotic, resulting in onerous contracting and operating conditions, which have severely inhibited all oil production in the country. Thus, any Venezuelan forecasts are problematic, because no one knows if and when the political situation will change.

We chose to be guardedly optimistic regarding future Venezuelan heavy oil production and assumed a crash program increment of heavy oil production of two million barrels per day after 20 years. That level could well be off the mark, and indeed, Venezuelan heavy oil production might even decline from where it is today, if mismanagement and political interference continues.

The incremental crash program heavy oil / oil sands production resulting from our assumptions is shown in Figure XI-3.

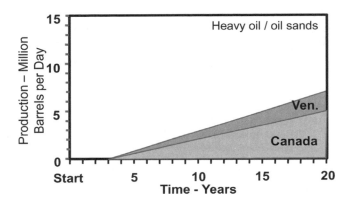

Figure XI-3. Estimated crash program of heavy oil and oil sands production from Canada and Venezuela. The estimates for Canada are aggressive, while the estimates for Venezuela assume a dramatic improvement in the country's political and investment environment.

D. Coal-To-Liquids as a Mitigation Pathway

Very clean hydrocarbon fuels can be made from coal. The favored coal-to-liquids (CTL) process involves heating coal to very high temperatures to break the coal down into various gases. Impurities are then removed, and liquid fuels and other products are created in the well-established Fisher-Tropsch (F-T) synthesis process, which produces a number of desirable liquids and solid hydrocarbons.

Gasification technologies have been steadily improved over the decades, and large numbers of gasifiers are in commercial operation around the world, many operating on coal. Advanced gas cleanup systems are currently utilized in refineries worldwide. F-T synthesis is also well developed and widely practiced. The liquid fuels resulting from coal gasification followed by F-T synthesis are of such high quality that they require little or no refining.

A number of CTL plants were built and operated in Germany during World War II. Subsequently, the Sasol Company in South Africa built its two, large-scale CTL plants under normal business conditions. A third facility -- Sasol Three -- was designed and constructed on a crash basis in response to a drop in oil imports resulting from the Iranian revolution in 1979. Sasol Three was completed in roughly three years, as essentially a duplicate of Sasol Two on the same site.

The Sasol Three experience represents the very best that might be accomplished in a twenty-first century crash program to build coal liquefaction plants. The South African government made a quick decision to replicate an existing plant near existing plants without the delays associated with site selection, environmental reviews, public comment periods, etc. Experienced personnel were readily available, and there were a number of manufacturers capable of providing the required heavy process vessels, pumps, and other auxiliary equipment.

Circumstances are different today. First, we are postulating many more than a single CTL plant being built at one time. Second, experienced design and construction personnel and supplier capabilities are now in short supply – a situation that will worsen with the advent of a worldwide crash program to build alternate fuels plants, which demand available, experienced personnel.

Our coal liquefaction crash program assumes that coal liquefaction plants would begin operation four years after a decision to proceed. We assume a plant size of 100,000 barrels per day of finished, refined products, and we assume that five such plants could be started each year. Where these coal liquefaction plants might be built is an open question. Candidate countries with large coal reserves are the most likely, which would include the U.S., Russia, China, India, and Australia.

When co-producing liquid fuels and electricity, overall costs are minimized and have been estimated to result in clean substitute fuels in the $60-80 per barrel range.

Our CTL crash program buildup is shown in Figure XI-4.

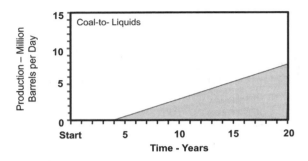

Figure XI-4. Worldwide fuel production from Coal-to-Liquids plants. CTLs begin operating after 4 years and reach a total production of 8 million barrels per day by year twenty.

E. Enhanced Oil Recovery as a Mitigation Pathway

Management of oil fields over their multi-decade lifetimes is influenced by an array of factors, including 1) expected future oil prices; 2) field production history, geology, and status; 3) cost and character of production-enhancing technologies; 4) the financial condition of the operator; 5) political and environmental circumstances; 6) an operator's other investment opportunities, etc.

When the decline of world oil production becomes widely recognized, oil prices will rise dramatically, and there is sure to be an explosion of interest in enhanced oil recovery (EOR). EOR is usually initiated after primary and secondary recovery have yielded most of what they can provide. Primary production is the process by which oil easily flows to the surface due to natural pressures. Secondary recovery involves the injection of water to force additional oil to the surface.

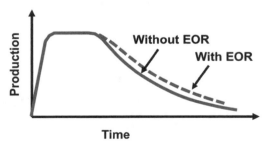

Figure XI-5. Increased production due to Enhanced Oil Recovery (EOR).

EOR has been in use for decades, particularly in the United States, where it currently is producing about 225,000 barrels per day of incremental oil. The process that likely has the largest worldwide potential involves the injection of carbon dioxide (CO_2) to add pressure and to act as a solvent to move remaining oil. As notionally shown in Figure XI-5, CO_2 flooding can increase oil recovery by roughly 7-15 percent of the original oil in place. However, the additional oil resulting from EOR does not flow quickly; it can take a decade or more to capture most of what's possible.

For many decades in the U.S., naturally occurring, geologically sourced CO_2 has been produced in Colorado and shipped via pipeline to West Texas and New Mexico for EOR. In the future, CO_2 can be collected from various CO_2 producing sources worldwide and shipped by pipeline and/or tanker to the large oil fields that are often not close to ready CO_2 sources – in the Middle East for instance. Such an effort will involve CO_2 collection, refrigeration, shipping, and injection into new wells in older oil fields. Considerable expense will be involved, including the cost of new tankers for shipping and new oil wells in older oil fields to optimize EOR production.

We assume that a crash EOR program will not begin to show enhanced production for 5 years after decisions to go ahead, because of the difficulties of procuring CO_2, moving it to candidate oil fields, drilling new wells, and allowing some time for increased oil flows to appear. We further assume that world oil production enhancement due to such a crash effort worldwide will increase world oil production by roughly 3 percent after 10 years. This is because even under a crash program, higher production increases seem unlikely, and not all oil fields are good candidates for EOR.

Our resulting added oil production estimate is shown in Figure XI-6.

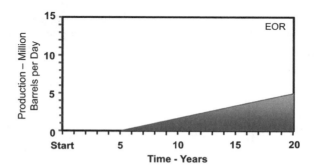

Figure XI-6. Worldwide Enhanced Oil Recovery. We assume that this increment is achieved primarily by CO_2 injection. It takes longer to begin incremental crash program EOR because of the difficulties of procuring and transporting CO_2 and drilling the new wells needed.

F. Gas-To-Liquids as a Mitigation Pathway

There are many large natural gas fields around the world that are isolated from gas-consuming markets. Significant quantities of this "stranded gas" are now being produced, the gas frozen into a liquid, and then transported to various markets in refrigerated, pressurized ships as liquefied natural gas (LNG). In a world short of liquid fuels, another possible option is to convert stranded natural gas into clean liquid fuels.

Gas-To-Liquids (GTL) processing is similar to CTL, and GTL technologies are well advanced. Of particular note is the Shell GTL project in Qatar, where the world's largest GTL plant was nearing completion at the end of 2009. When fully operational, it is expected to produce 320,000 barrels per day of liquid fuels and 120,000 barrels per day of Natural Gas Liquids (NGLs) and ethane, a light hydrocarbon gas. Other smaller–scale GTL plants are in operation, providing a solid basis for moving aggressively to implement the technology on a massive scale in order to provide a large source of liquid fuels.

Developing large GTL projects involves a combination of drilling wells to produce natural gas, along with the construction of large conversion facilities, which resemble a very large oil refinery.

In a crash program to mitigate growing shortages of oil, GTL plants might be built in a number of countries that have large uncommitted reserves of natural gas. Once operational, GTL liquid fuels can be moved to markets around the world by conventional oil product tankers.

For our estimate of worldwide GTL crash program contributions, we assumed a startup delay of three years before new GTL plants might come into operation. Based on our reading of the situation, we assumed that a crash program might yield the 1.0 MM bpd in 5 years. The resultant contributions are shown in Figure XI-7.

> *"Gas-To-Liquids (GTL) processing is similar to CTL, and GTL technologies are well advanced."*

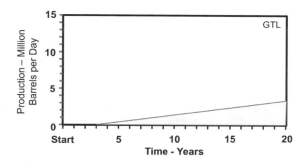

Figure XI-7. Worldwide GTL liquid fuels production as a result of a crash program.

G. Results of Crash Program Physical Mitigation

1. Review of Our Purpose

The purpose of the preceding considerations was to identify the technologies most likely to successfully mitigate world oil production decline and to roughly estimate the possible contributions of each. The result provides an estimate of the upper limit on what might be accomplished -- the best possible scenario. Our estimates necessarily involve making a number of judgments on what might be accomplished with each option. In developing the foregoing, we used available literature, actual data, and our experience. Again, we remind you that the problem we are considering is the physical mitigation of future oil shortages – shortages that will increase over time. All options capable of large-scale contributions will be needed, because the task is much bigger than any single solution – "there is no silver bullet." To repeat, the emphasis is on liquid fuel savings and production, because the problem is a liquids problem, not simply energy.

> *"To repeat, the emphasis is on liquid fuel savings and production, because the problem is a liquids problem, not simply energy."*

2. Assumption on Future World Oil Production

As we said, the timing of the onset of the decline of world oil production cannot be predicted with certainty. Our model projects that the recent fluctuating world oil production plateau will continue until the point of decline.

It is our belief that the break point will occur within the next five years, after which production will decline annually. We considered two annual decline rates: 2% and 4%. The resulting world oil production pattern is shown in Figure XI-8

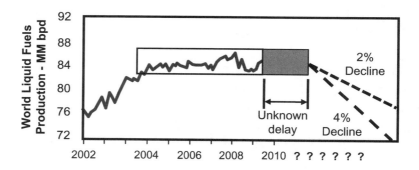

Figure XI-8. Historical world oil production and our forecast for the future. The long fluctuating plateau period started in mid 2004. We assume it will continue for an unknown number of years – five at the most—after which production decline will occur at 2% or 4%.

It is instructive to note how the world oil market reacted to the recent world oil production plateau, during a period of growing world GDP. The experience is not surprising. Between mid 2004 and mid 2008, world GDP expanded at an average rate of roughly 3.6% per year, while world oil production remained essentially flat. Because of the tight coupling between world oil supply and world GDP, the stagnant supply situation coupled with a growing GDP could only be accommodated by increasing oil prices. During the four-year period, oil prices increased by roughly a factor of four until the "Great Recession" abruptly changed the world economy.

If world oil production were to stay within the current fluctuating production band, as we have postulated, a significant oil price escalation is almost certain when world GDP expands after the current recession ends. When world oil production goes into decline, oil prices will increase even more dramatically, putting a major drag on world GDP.

3. The Sum Total of Crash Program Physical Mitigation

We are now in a position to add our five contributions to generate an estimate of what an idealized crash program of physical mitigation might provide. The results are shown in Figure XI-9.

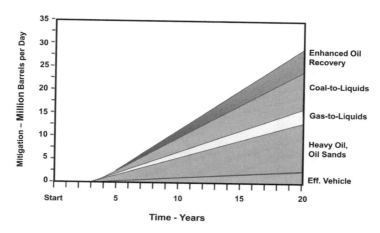

Figure XI-9. The sum total of crash program physical mitigation estimates. Contributions begin three years after a decision to go ahead and grow to roughly 30 million barrels per day after 20 years.

4. Crash Program Mitigation for 2% & 4% Decline Rates

Earlier, we considered two likely world oil production decline rates of 2% and 4%. We now ask how our worldwide crash program of physical mitigation might soften the oil shortages in those two cases. The results from Figure XI-9 applied to 2% and 4% decline rates are shown in Figures XI-10 and XI-11.

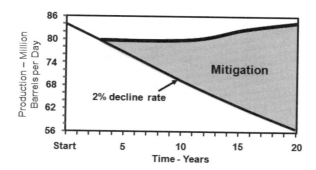

Figures XI-10. The impact of crash program mitigation on oil supply when world oil production declines at 2% per year. In this situation overall production declines modestly and remains relatively flat at a lower level for more than a decade.

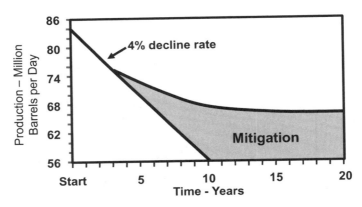

Figures XI-11. The impact of crash program mitigation on oil supply when word oil production declines at 4% per year. In this situation overall production declines dramatically and becomes relatively flat at a dramatically lower level after roughly a decade.

In the 2% decline rate case, the resulting world oil shortages would be clearly damaging, especially considering the attendant oil price increases. In the 4% case, economic damage would be severe.

5. What will a realistic worldwide crash program look like?

Our ideal crash program was postulated to start instantaneously, meaning that business-as-usual would be instantly followed by full-blown implementation. This assumption is useful from the point of view of understanding the best that is possible, but in the real world, such huge changes cannot happen that quickly for a number of reasons:

1. Governments will have to facilitate the rapid permitting of facilities, eliminate legal challenges to moving ahead quickly, provide incentives for industry to act aggressively, and provide financial protection for industries that undertake large mitigation risks. It is virtually certain that governments will not move quickly on such matters, because of the need to change priorities and to understand the practical options.

> *"Our ideal crash program was postulated to start instantaneously, meaning that business-as-usual would be instantly followed by full-blown implementation."*

2. There is such strong sentiment among environmentalists and people concerned about climate change that there will be huge pushback until it is obvious that the decline of world oil production is causing severe hardship.

3. Industry does not now have the capacity, infrastructure, and trained personnel to dramatically expand into a full-blown crash program overnight. Where industrial capacity does exist, it is often located in other countries, where local priorities would almost certainly restrict providing services or hardware to other countries.

4. The efforts envisioned will be very expensive, and the large financing requirements are unlikely to be available quickly.

5. Not all countries with the resources to undertake meaningful physical mitigation may do so quickly.

6. Any effort by governments to pick winners and losers would almost certainly be disastrous, based on recent history.

7. Finally, there is a tendency of governments to want to exert control over big programs. If excessive, government control could severely damage what only private industry can do effectively.

Thus, a realistic implementation pathway is certain to be slower than we postulated, more like what is shown in Figure XI-12. The result is that oil shortages will almost certainly be larger than in our idealized model.

Why not assume a more likely, slower startup crash program pathway? Our answer is that we were seeking the best response possible, which would be an overnight scale-up. In addition, we know that an analysis of more likely startup pathways is a complicated job involving a number of subjective assumptions.

From these considerations, it is clear that the mitigation results shown in Figures XI-10 and XI-11 are certain to be slowed by at least a year, if not longer, resulting in even more serious oil shortages.

According to our model, the bottom line is that the world is facing significant oil shortages after the onset of the decline in world oil production. If the decline rate is a relatively modest 2%, the shortages will be less onerous than if the decline rate were 4%.

> ## *"The bottom line is that the world is facing significant oil shortages after the onset of the decline in world oil production."*

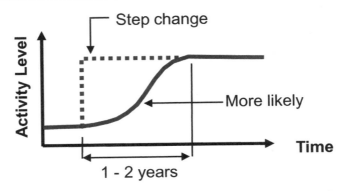

Figure XI-12. Comparison of a step change pathway with a more likely pathway to change.

6. What about global warming?

Implementing most of the foregoing technologies will result in the emission of significant amounts of carbon dioxide, believed by many to be a major factor in global warming. Our thinking is that the first priority of populations and responsible governments is the immediate welfare of its citizens. The economic damage resulting from annually increasing oil shortages is certain to be devastating, so the first priority of nations will be economic stabilization, which will be totally dependent on physical mitigation.

Later in the book we discuss various aspects of global climate change.

> ## *"The first priority of nations will be economic stabilization, which will be totally dependent on physical mitigation."*

XII. Let's save our way out of impending oil shortages.

A. Introduction

Increasing the energy efficiency of vehicles, machinery, and equipment fueled by oil products makes excellent sense, as long as the related costs are reasonable and owners can afford the changes. In some cases, retrofits are conceivable, but in most cases, implementing significant energy efficiency means buying something new. Another difficulty is that vehicles, machinery, and equipment tend to be large and expensive, unlike computers and electronics, where unit sizes are small, facilitating rapid changes.

In the following, we use the U.S. as our basis for discussion, because good information and data were readily available. However, our most important conclusions are broadly applicable to other countries.

The story is as follows: 1) The world's oil consuming vehicles, machinery, and equipment have long lives and represent huge investments; 2) While replacements with more liquid fuel-efficient or alternate energy technologies are possible, change cannot happen quickly because purchases are typically very expensive; 3) When world oil production goes into decline, a serious recession will ensue, which means that many vehicle, machinery, and equipment owners will have difficulty affording replacements. [81]

B. Transportation Machinery

Because transportation is such a large consumer of oil products, its consideration is significant in the context of impending oil shortages. We examined the U.S. for 2008, because it was a year not distorted by recession. Summary data are shown in Table XII-1. [82]

In 2008 automobiles represented the largest single oil product consuming capital stock, with 136 million autos consuming 4.9 million barrels per day, or 25 percent of total U.S. liquid fuels consumption. A recent U.S. Department of Transportation study showed that one half of recent model year cars will remain on the road over 13 years later. If applied to model year 2010, over one-half of the vehicles made in 2010 will still be on the road in 2023, and a measurable number will still be on the road in 2036. At normal replacement rates, consumers would spend an estimated $2 trillion (2008 dollars) over

"In most cases, implementing significant energy efficiency means buying something new."

the next 10-15 years to replace just one-half of the automobiles now on the road.

Table XII-1. 2008 U.S. Transportation Fleet Profiles.

Characteristic	Autos	Light Trucks	Heavy Trucks	Aircraft
Fleet Size (Millions)	136	102	9	0.24
Annual Purchases (Millions)	7.7	8.5	0.45	0.008
Rough Cost of the Fleets (Billions)	700	1,000	200	400
Share of Oil Consumption	25%	20%	13%	7%
Median Lifetime of Vehicles	13	14	18	22

A similar situation exists with light trucks, which includes Sport Utility Vehicles (SUVs), vans, and pick-up trucks. In 2008, this class of vehicle consumed 4.0 million barrels of oil products per day, accounting for 20 percent of total oil consumption. The median lifetime of light trucks was over 14 years. At current replacement rates, one-half of the 102-million light trucks existing in 2008 will be replaced in the next 11-16 years at a cost approaching $1.5 trillion.

One of us (RHB) served on a 2010 National Academies study that assessed technologies and approaches to reducing the fuel consumption of medium and heavy-duty vehicles. [84] Nine million heavy trucks, including buses, highway trucks, and off-highway trucks, represent the third largest consumer of oil products in 2008 at 2.6 million barrels per day, 13 percent of total consumption. The median lifetime of this equipment is 18 years. At normal replacement levels, one-half of the heavy truck stock will be replaced by businesses in the next 15-20 years at a cost of $1.5 trillion.

The fourth-largest consumer of oil products is airplanes, which consume the equivalent of 1.3 million barrels per day, representing seven percent of U.S. consumption in 2008. The 8,500 aircraft used by scheduled airlines and 232,000 privately owned aircraft have a median lifetime of 22 years. Applying recent rates, one-half of the aircraft fleet will be replaced over the next 20-25 years at a cost of almost $1 trillion.

C. Other Oil Usage

The four transportation fleet categories cover most transportation modes and represented 65 percent of the consumption of oil products in the U.S. in 2008. The largest remaining oil-consuming capital stock resides in the industrial sector, which is diverse, making it difficult to identify potential efficiency options or potential technology advancements. The largest oil-consuming industries include the chemical, lumber and wood, paper products, and the petroleum industry itself. Industrial use of oil includes process heat, power, feedstock, general heating, and lubrication. Finally, it is important to note that industry has long been concerned about operating costs, so when more efficient, cost-effective equipment has become available, it has typically been brought into use on a timely basis. While additional efficiency opportunities undoubtedly exist, we do not believe that there are huge gains or large energy source changes immediately available for U.S. industry.

D. Efficiency Improvements in Autos and Light Duty Trucks (Light Duty Vehicles)

In the previous chapter, we discussed options for liquid fuel savings in light duty vehicles. In a landmark study on a number of light duty vehicle options by the National Academies on which one of us served (RLH), future efficiency gains in a number of optional vehicle types were estimated. [85] The Academies estimated the amount of gasoline that might be saved in the U.S. between 2010 and 2050 for various vehicle power options. The results for two types of plug-in hybrid electric vehicles (PHEVs) are shown in Figure XII-1.

As we noted, the assumptions underlying these estimates were those in a recent U.S. DOE Energy Information Administration (EIA) forecast to the year 2030, which the Academies committee extrapolated to the year 2050. The EIA forecast assumed business-as-usual for the period and did not include any consideration of the impending decline in world oil production, which would dramatically change most aspects of their forecasts.

A fair question is: Why show the Academies estimates, if we do not believe the underlying EIA assumptions? The answer is that the Academies numbers illustrate two fundamentals, which we believe are important and universal:

1) If the U.S. economy were to develop without incident over the next 40 years and if technology were to evolve as it has, U.S. light duty vehicle consumption of gasoline would not change dramatically for two decades, after which it would trend upwards. In other words, under the best of normal conditions, significant change does not occur rapidly.

2) Under optimistic assumptions regarding market penetration, the advent of plug-in hybrid vehicles (PHEVs) would not have a noticeable impact on oil consumption for more than a decade, after which a significant decrease in consumption of gasoline would accrue. Under the best business-as-usual conditions, significant change does not occur rapidly.

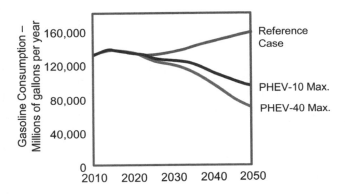

Figure XII-1. U.S. gasoline consumption to the year 2050 in a reference case and with the optimistic deployment of plug-in hybrid electric vehicles (PHEVs).

E. Concluding Remarks

The rapid replacement of less-efficient oil-product consuming vehicles, machinery, and equipment with more efficient replacements would help mitigate the economic impacts of rising oil shortages and escalating oil prices caused by the decline of world oil production. However, replacement rates normally imply once every 10-20 years and overall costs of trillions of dollars for the U.S. alone. For the rest of the world, the costs would be many times U.S. costs.

While the 2009 U.S. federal "cash-for-clunkers" program accelerated the replacement of over 600,000 autos, it did so at a $3 billion price tag that was not sustained because of cost and is not likely to be viable in the future in a country already facing massive federal deficits and that will find itself in a deepening recession caused by oil shortages. We thus find it difficult to conceive of an affordable government-sponsored crash program to accelerate normal replacement schedules to introduce higher energy efficiency and alternate energy technologies into the transportation sector. Significant improvements are inherently time-consuming, because of the size of the fleets of vehicles, equipment, and machinery and the costs of replacements, especially during recessionary times. Another complicating factor is that in times of recession, most consumers will not be able to afford to replace their vehicles – especially since declining oil production will reduce the trade-in values of their existing, less fuel efficient vehicles. Our conclusion is that we cannot simply save our way out of impending oil shortages.

> *"Significant improvements are inherently time-consuming, because of the size of the fleets of vehicles, equipment, and machinery and the costs of replacements, especially during recessionary times."*

XIII. Criteria for Judging Energy Processes: EROEI and LFROI

A. Introduction

Individuals and businesses invest money looking to earn a return on their investments. An end result *greater than one* represents a gain, and an end result of *less than one* means money was lost. A very good investment provides a return much greater than one.

A similar criterion applies in energy. To produce useful energy, be it liquid fuels or electricity, we must invest energy. In the realm of energy production, the term that is often used is called Energy Return on Energy Invested – EROEI, sometimes EROI. EROEI is an increasingly important concept in energy economics and market analysis. For that reason, it is worth some time to understand the concept and its cousin, Liquid Fuel Return on Investment – LFROI.

> *"To produce useful energy, be it liquid fuels or electricity, we must invest energy. "*

B. The Concept

For any energy production process, its EROEI is defined as the useful energy produced by the process divided by the energy that humans must invest to produce it. As an example, consider the production of oil from a new well. In simple terms, people and heavy equipment must be transported to the drill site. One or more wells must be drilled, which requires a good deal of energy; in the process, drilling fluids must be pumped into the well or wells, which requires energy; the well or wells must then be hooked up to new or existing pipes and pumps are needed to move the oil to a processing center, etc. Some of these energy requirements are relatively easy to estimate, but some are difficult.

> *"For any energy production process, its EROEI is defined as the useful energy produced by the process divided by the energy that humans must invest to produce it."*

An EROEI of much greater than one means that much more useful energy is obtained than the invested human-applied energy. EROEI is a valuable tool for assessing the potential of various types of energy production processes.

C. Energy Returns on Energy Invested for Different Energy Sources

"Studies conducted on the EROEI for different energy processes often yield varying estimates."

The EROEI for oil production in the 1930s has been estimated to have been as high as 100:1. [86] That energy return provided huge dividends for society. As time went on and it became increasingly more difficult to produce oil, the EROEI for oil production declined. Recently, the EROEI for oil production worldwide was estimated to be about 14 to 1, meaning it took the energy equivalent of 1 barrel to produce 14 barrels of oil. That is still high but clearly not as good as it was.

As we discuss below, estimating EROEI is very difficult, so results are necessarily approximations. Indeed, studies conducted on the EROEI for different energy processes often yield varying estimates.

Consider some recent estimates for a number of liquid fuel and electric power production processes, as shown in Figure XIII-1. [87]

- As noted, the EROEI for oil production during the 1930s was very high, and while it is now lower, it is still very attractive.

- The EROEI for coal production, hydroelectric power, and wind electricity is also high. There are huge coal resources yet to be produced, but new hydroelectric opportunities are generally limited. The EROEI of wind is attractive but it has a serious intermittency problem, which severely limits its practical value.

- The EROEI for biomass liquid fuels options such as ethanol and biodiesel is low and may actually be less than unity in some cases, meaning that more energy is consumed to produce a gallon of product than was invested by human efforts. In particular, corn-ethanol has turned out to be unattractive from an EROEI point of view.

D. Oil EROEI

In the oil industry, early oil fields were found at relatively shallow depths, so the energy required for drilling and production was relatively low. The result was a very significant energy payback, yielding large amounts of useable energy that fueled world economic expansion. The early days of oil production yielded huge energy returns on investment, allowing society to advance dramatically.

Over the years, as oil producers had to drill deeper into more difficult rock formations, the EROEI for oil production declined. The paybacks today are much lower than in the early years, as displayed in Figure XIII-2. [88]

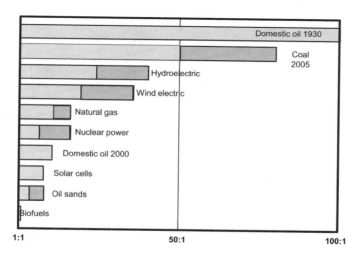

Figure XIII-1. The energy return on investment for a number of energy sources. Light grey segments of the bars indicate uncertainty in the estimates. Some uncertainties are seen to be very large.

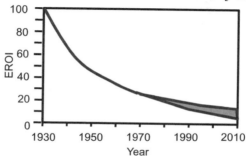

Figure XIII-2. Estimates of EROEI for oil and gas production from the 1930s to 2010.

The declining trend of EROEI in the oil industry is troubling from a societal point of view. As Charlie Hall points out, "humans use high quality, low-cost resources before low quality, high cost resources." [89] That means that the energy payback in the future will be still lower, a fact that needs to be factored into comprehensive energy planning.

E. Conceptual Difficulties in Estimating EROEI

While the concept is useful, accurately estimating the EROEI for different energy production processes is extremely complicated, resulting in differing estimates for various energy production technologies. There exists no formal standard on the measures to calculate EROEI. In addition, the form of the energy input is often different from the output. For example, coal is sometimes used in the production of corn-ethanol, a fact that tends to muddy the supposed purity that some ascribe to that corn-ethanol.

A question revolves around how far back in the energy supply chain the analysis needs to go. For example, if steel is being used to drill for oil or to construct a nuclear power plant, should the energy input of the steel be taken into account? Should the energy used in building the factory used to construct the steel be taken into account and amortized? Should the energy used to construct the roads used to transport the goods be taken into account? [90] These are complicated, some would say impossible questions that defy simple answers. Nevertheless, the casual observer gains insights into the ubiquitous nature of energy just by considering these questions.

Another problem is that EROEI does not take into account the factor of time. Energy invested in creating solar cells is difficult to compare to a power source like coal, because the energy return from the solar cells accrues very slowly over maybe two decades, while coal tends to be used relatively soon after it is mined. Use of a reasonable discount rate would greatly reduce the solar EROEI. If EROEI is to become a more accurate measure, specific, detailed ground rules would need to be developed. Is the time and effort required to develop that detail worth the effort, or will approximations suffice? Rough estimates already tell us useful things about energy options.

Finally, there exist no formal accounting rules for the consideration of waste products created in the production of useful energy. For example, waste products generated in the production of ethanol make its EROEI calculation difficult. It is not simple to justify how corn-ethanol waste should be counted, since it is used for animal feed, which is a positive.

F. Liquid Fuels Return on Investment (LFROI)

"Liquid Fuels Return on Investment (LFROI)" refers to the amount of

> **"Liquid Fuels Return on Investment refers to the amount of liquid fuels produced by a liquid fuel production process divided by the liquid fuels required to produce them."**

liquid fuels produced by a liquid fuel production process divided by the liquid fuels required to produce them. This concept is important on the production side of the impending decline of world oil production, since it would make no sense to consume a large quantity of liquid fuels in order to produce a relatively small incremental quantity. LFROI must be considerably greater than unity, if there is to be a rapid and efficient rise in liquid fuels production to offset increasing world oil shortages; in fact the higher the LFROI for a process, the more attractive it is for the mitigation of declining world oil production.

We know of no good analyses of LFROIs for important liquid fuel production technologies. Instead, analysts have focused on EROEI.

Based on our familiarity with the technologies that are likely to be involved on the production side of the mitigation of declining oil production, we believe that LFROIs of 10-20 are likely for the following sources of liquid fuels:

- Conventional oil fields
- Deepwater oil & Arctic oil
- Enhanced Oil Recovery
- Oil Sands
- Heavy Oil
- Gas-To-Liquids
- Coal-To-Liquids
- Shale oil

Based on the LFROI criteria and our judgments, we believe these options to be attractive for oil shortage mitigation.

One class of liquid fuel production options does not measure up well when judged on the LFROI criteria: Some of the Biomass-To-Liquids (BTL) technologies. The reason is that most BTL crops are grown over very large land areas, requiring the expenditure of large amounts of liquid fuels to plant, harvest, collect, and transport the biomass to a processing center. A rule-of-thumb in land-based biomass systems is that they cannot profitably be larger than about 50 miles in radius, because of the increasing liquid fuel requirements of going further out.

> **"Taxpayers are providing subsidies for corn ethanol that yields an expensive, low energy content, low-LFROI fuel."**

The reasoning involves simple geometry. Land-based biomass is spread out over a large area, because plant density is always relatively low, even for the densest of crops. In all cases that we know of, biomass must be transported to a central location for processing, since taking the processing equipment to the biomass is impractical. Biomass that is close to the processing facility does not have far to travel, so close-in biomass does not require very much liquid fuel for gathering and transport, because the area close in is small. It is the land areas that are further away from the central processing plant where the greatest quantities of biomass are located; a swath of biomass along a distant circumference is going to contain much more biomass than one at a small circumference, because there's much more surface area. But far-away biomass must, by definition, be transported further to the processing facility, and the machinery to do that runs on liquid fuels, so more liquid fuels are consumed and less net fuel is available after the biomass is processed into liquid fuels. [91]

In the case of corn-ethanol, there are additional liquid fuel expenditures for soil tilling and watering, often provided by liquid fuel powered pumps. We believe that the corn-ethanol mandates are ill advised, when viewed from through the lenses of EROEI and LFROI. Basically, taxpayers are providing subsidies for corn-ethanol that yields an expensive, low energy content, low-LFROI fuel.

G. Some conclusions

Liquid Fuel Return on Investment (LFROI) is extremely important in selecting technologies to provide liquid fuels. If a technology has a low EROEI but a high LFROI, it may be an attractive option during the difficult period where all viable, high LFROI liquid fuel production options will be needed to mitigate world liquid fuel shortages.

Longer term, Energy Return on Energy Invested (EROEI) is an essential measure in planning for future energy systems. Both criteria require greater consideration than has heretofore been provided.

> **"Liquid Fuel Return on Investment is extremely important in selecting technologies to provide liquid fuels."**

XIV. Other Transportation Fuel Options

A. Introduction

The greatest concerns related to the decline in world oil production are the impacts on transportation, which depends almost entirely on liquid fuels. In our chapter on physical mitigation, we described technologies that are ready for widespread implementation to help mitigate growing world oil shortages and associated very high oil prices.

Other physical options are also conceivable, including natural gas, biomass, shale oil, and hydrogen. In this chapter, we consider each.

B. Natural Gas

1. Introduction

Natural gas is used for residential, commercial, and industrial building heating, cooking and drying, electric power generation, feedstock for the production of a number of chemicals, and fueling some transportation vehicles. Natural gas burns relatively cleanly and has been relatively inexpensive in many parts of the world.

Natural gas is typically transported to end-users by pipelines, typically underground. Its use has expanded in recent decades, particularly for electric power generation, because of its relatively low emissions, and the fact that natural gas-powered electric power plants can be built rapidly at relatively low cost.

Natural gas is currently an important energy workhorse in many countries, and the outlook for its future use is very promising. Because the supply picture has changed dramatically in the last few years, natural gas fueling of more of the transportation system is worthy of serious consideration. In the following, we provide some background, reasons for the supply optimism, and some considerations related to its use as a transportation fuel.

> *"Natural gas is currently an important energy workhorse in many countries, and the outlook for its future use is very promising."*

2. What It Is

Natural gas is a fossil fuel, primarily methane, a light hydrocarbon composed of one carbon atom surrounded by four hydrogen atoms. Because methane has so many hydrogen atoms per carbon atom, burning of natural gas emits less carbon dioxide per unit of energy than coal, which has fewer hydrogen atoms per carbon atom.

When natural gas is produced, it is almost always accompanied by Natural Gas Liquids (NGLs), composed of heavier -- but still light hydrocarbons known as ethane, propane, and butane, which are liquid under normal conditions. When natural gas is produced, it contains impurities such as nitrogen, water vapor, and hydrogen sulfide, which are routinely and easily extracted, leaving a relatively clean fuel.

3. Where It Comes From

Natural gas exists in the earth in different geological settings. First and foremost, it is found in underground reservoirs, either alone or in conjunction with oil. Natural gas reservoirs are distributed in various locations around the world, often in regions where oil is found. Like oil, natural gas is extracted by drilling wells, which provide paths for the gas to flow to facilities on the surface that separate out natural gas liquids and impurities, after which the clean gas flows to end users.

Large quantities of natural gas are found in places that are remote from markets. For that reason that gas is called "stranded," because it cannot be economically pipelined to markets. When stranded natural gas is developed, it is typically pipelined to shipping terminals, where it is liquefied by cooling it to extremely low temperatures, which facilitates its transport via special tankers. At receiving terminals, the liquefied natural gas is off-loaded and heated to its gaseous state so it can be fed into pipelines for transport to end-users.

Like oil, the world's natural gas resource is finite, so it will be exhausted at some future date. However, the remaining resource is significant. Natural gas is also found in coal seams and in so-called gas shales. Methane in coal mines is a safety hazard, so special care is taken in its handling. In deep coal beds, wells drilled into the coal can extract significant quantities of natural gas, even though the coal is not mined.

> *"Natural gas exists in the earth in different geological settings."*

Shale gas is natural gas trapped in shale rock. In recent years it has begun to be widely exploited as a relatively new source of so-called unconventional natural gas, creating optimism regarding future natural gas supply, because gas shales are found in so many places around the world.

Two other sources of natural gas are worthy of mention. Biogas is methane created in landfills, swamps, marshes, sewage sludge, and manure, as a result of the decay of organic matter. Nature has been creating methane in this way for millennia. Biogas normally finds its way into the atmosphere, where it acts as a greenhouse gas.

In landfills, methane is trapped by overburden and can be removed for beneficial use via relatively inexpensive, shallow wells. This method of methane capture is practiced in many places and transforms a nuisance into a useful source of small-scale supplemental energy.

Methane can be produced by processing coal. Related processes typically involve heating coal to high temperatures, where it breaks down into various molecules, including methane. Many decades ago, coal heating was carried out in simple processing facilities, creating what was called "Town Gas," used in home and building heating. Town gas was typically laden with impurities, which meant that its use created health and environmental problems. Today, there are sophisticated processes for manufacturing a very clean natural gas from coal, but these processes are not widely practiced, because natural gas from other sources is usually less expensive.

Finally, it is worth noting that huge amounts of natural gas occur in the form of gas hydrates, which are ice-like solids in which methane is trapped in a microscopic cage of water molecules. Practical production of gas from hydrates is not now commercial.

4. Natural Gas Use in the U.S.

Since we have good data, we consider the U.S. situation, recognizing that many things that apply in the U.S. apply elsewhere in the world.

Natural gas accounts for over 20% of total U.S. energy consumption. About 35% of natural gas is used in the residential sector, 25% in the industrial sector, 20% for electrical power generation, and about 20% for commercial uses. [92]

In residences, natural gas accounts for about 50% of space and water heating. In industry, natural gas is also used for space and water heating, as well as for incineration, metals processing, drying, waste treatment, glass melting, food processing, fueling boilers, and powering large cooling systems. Natural gas is also used as a feedstock for the manufacturing of a number of chemicals, fertilizers, and other products. Processing natural gas with steam can produce hydrogen, which is in turn used in oil and chemical refineries.

> *"In electric power generation, natural gas has become increasingly popular, because gas turbine power stations have relatively small environmental impacts."*

In electric power generation, natural gas has become increasingly popular, because gas turbine power stations have relatively small environmental impacts, and they can be relatively easily permitted and rapidly constructed. They can produce relatively low cost electric power, if natural gas prices are low. Electric power from natural gas generating plants can be easily and rapidly cycled up and down at reasonable costs, which is of particular value to compliment the down cycles in wind and solar power generation.

Commercial uses of natural gas parallel residential uses, e.g., space and water heating, cooling, food preparation, on-site small scale electric power generation, etc.

5. Natural Gas From Shale

Shale gas is natural gas trapped in shale rock, which is fine grained, brittle, laminated, and not very porous. Natural gas cannot easily move between the tiny pores in shale rock, because of the low connectedness of the rock pores, which is called permeability. Shale gas exists in many countries and regions, including Australia, Canada, China, Europe, India, and the United States.

> *"Shale gas exists in many countries and regions, including Australia, Canada, China, Europe, India, and the United States."*

It has long been known that there are very large amounts of natural gas in shale formations, and shale gas has been produced in small quantities for over a century. Because the large-scale extraction of natural gas from shales has been inherently more expensive than extraction from typical natural gas reservoirs, shale gas has been of little interest in years past. It is only recently, when the easy sources of natural gas production have gone into decline in many regions that shale gas has emerged as being attractive.

The techniques for producing shale gas include vertical and horizontal drilling and fracturing. In simple terms, the sequence involves drilling vertically down into shale rock – often many thousands of feet underground -- and then turning the drill bit horizontally and drilling very long distances out into the shale rock – sometimes as far as 10,000 feet. After drilling such a long, L-shaped well, fracturing is used to crack the shale rock at frequent intervals along the horizontal portion of the well. The resulting fractures are held open with propping materials, providing channels that natural gas can flow to. The fractures are frequent enough that much of the gas has a relatively short distance to travel through the shale to reach an open channel, which connects to the horizontal well, through which it can flow to the vertical part of the well, and then up to the surface, where it is processed for beneficial use. Horizontal drilling and fracturing for shale gas production represent advanced developments of technologies that have been in use in the oil and gas industry for decades.

> *"Shale gas production from newly fractured wells is very high during the first year of operation and declines dramatically thereafter."*

Because it is so difficult for natural gas to flow through low permeability shale rock, shale gas production from newly fractured wells is very high during the first year of operation and declines dramatically thereafter. This is the result of gas close to fractures being able to exit quickly, while gas further away from fractures takes longer to get to fractures. Production from shale gas wells can decline 50 percent or more in the first year, which is dramatically faster than the flows of typical natural gas wells, which flow at high levels for long periods of time. Shale gas wells are kept in operation for many years, but the greatest production occurs in the first few years. The situation is a little like a person taking a deep breath and exhaling hard – There is a sudden burst of air followed by less and less air as lungs collapse.

> *"Maintaining high overall levels of shale gas production requires the continuous drilling of new shale gas wells."*

In practical terms, this pattern of shale gas production

means that maintaining high overall levels of shale gas production requires the continuous drilling of new shale gas wells. In principle, continuous drilling is possible, but there is one sticky problem. Natural gas prices are typically volatile, so shale gas operators may not make enough money during periods of low natural gas prices to maintain continuous drilling. A prolonged period of low natural gas prices could cause shale gas drilling to slack off, which would lead to a relatively rapid drop in natural gas supply, leading to gas shortages, which would mean much higher gas prices, which would then stimulate operators to restart shale gas drilling. The resulting volatility could lead to aggravated boom and bust cycles, which would not be in anyone's interest – consumers, operators, shale gas workers, etc. Further, it could make natural gas a less attractive option for electric power generation. How often and how severe this destructive volatility sequence might develop remains to be seen.

> *"The issues associated with the massive development of shale gas are cost, water use and disposal, environmental impacts, and the need for continuous, uninterrupted development."*

The issues associated with the massive development of shale gas are cost, water use and disposal, environmental impacts, and the need for continuous, uninterrupted development.

Thousands of shale gas wells were being drilled in the U.S., as of the beginning of 2010, when we were writing this book. Natural gas prices were roughly $4 per million Btu (British Thermal Unit), the standard industry measure, which is based on energy content. Some have argued that this price is marginal for shale gas development, but the fact that vigorous drilling still occurred suggests that $4 may be adequate, or maybe it is not, and shale gas drilling will slack off in the near future. [93]

Because shale gas development requires large-scale fracturing, large amounts of water are needed to carry fracture chemicals and propping materials into the shale formations. But large quantities of water are not always readily available near drilling sites. A related problem is that fracturing water returns to the surface after fracturing is complete and because the return water is contaminated by fracture chemicals and impurities present in the shale rock, water cleanup is important for environmental protection.

There have been reports of fracture fluids contaminating water in underground aquifers, which supply water for human and animal consumption. These reports have upset the public and have already led to bans of shale gas development in some regions. For example, in April 2010 New York State imposed restrictions on shale gas drilling in watersheds for the cities of New York and Syracuse, citing concerns over drinking water safety. For these cities, the restrictions amount to a "de facto ban" on hydraulic fracturing, and environmentalists are seeking a blanket ban on shale gas drilling in all watersheds. [94]

There have also been calls for federal regulation of fracturing, instead of at the state level, where regulation currently exists. State regulation of fracturing in conventional oil and gas fields seems to have worked well over many decades, although some argue that point. Washington involvement could introduce political pressure into fracturing regulation, something that operators greatly fear.

If the federal government becomes involved in fracturing regulation, the industry belief is that the development of shale gas is almost certain to be negatively impacted. This point is illustrated by the recent ExxonMobil $41 billion acquisition of XTO Energy, a major developer of shale gas. Completion of the deal hinged on the federal government not getting involved in fracturing regulation.

Shale gas production may also have other environmental problems, and a complete consideration of all emissions may render it less attractive from a greenhouse gas point of view. For instance, a recent study compared the total greenhouse gas emissions of shale gas and estimated them to be more than twice the emissions from the combustion of natural gas sourced conventionally. [95] When total emissions are considered, shale gas and coal from mountaintop removal may have similar releases. The study concluded, "Society should be wary of claims that natural gas is a desirable fuel in terms of the consequences on global warming."

> *"If shale gas is to become a mainstay of natural gas supply, it may be necessary for governments to place a floor on natural gas prices to avoid boom-and-bust cycles."*

Finally, as noted, continuous, uninterrupted development of shale gas is necessary because the natural gas production from shale wells declines rapidly after the first year of production. If shale gas is to become a mainstay of natural gas supply, it may be necessary for governments to place a floor on natural gas prices to avoid boom-and-bust cycles. When governments get involved in markets, political pressures have been known to upset pragmatic decision-making. Is government involvement justified, desirable, or practical? There are difficult questions to be resolved.

6. Methane Hydrates

As noted earlier, methane hydrates are ice-like solids wherein methane is trapped in a microscopic cage of water molecules. These hydrates exist in many locations worldwide in cold regions, such as the arctic, and underwater, where temperatures are higher but where high pressures maintain the hydrates.

The hydrate resource is enormous but not estimated with great accuracy. However, "enormous" is enough to capture people's interest and to justify significant research on how methane hydrates might beneficially be produced.

A recent study by the U.S. National Academies stated, "… the environmentally and economically sustainable production of methane from methane hydrate … has not yet been achieved. Complex scientific challenges, which may require the development of new technologies, remain before methane from methane hydrate can be realized as an energy resource." [96]

Three primary methane hydrate production concepts have been proposed: 1) depressurization; 2) heating; and 3) chemical stimulation. These approaches involve liberating methane from its water molecule cage, and therein lies a major problem. To our knowledge, most methane hydrate formations are not located below an impermeable cap rock or other formation that can effectively contain the methane that is freed up by human activity. Recall that cap rocks are an essential element in oil and gas reservoirs; cap rocks act like inverted bowls and have held oil and gas in underground reservoirs for millions of years.

Without cap rocks or other containment, methane released from hydrates will go where it wants. Since it is a light gas, methane tends to move vertically away from its former place of residence, eventually escaping into the atmosphere. That reality does not facilitate its easy capture; it could create a fire hazard, and it

> **"Without cap rocks or other containment, methane released from hydrates will go where it wants."**

could enhance the greenhouse effect. The absence of a cap rock or other containment structure thus represents a major impediment to the practical capture of methane hydrates. Nevertheless, because the methane hydrate resource is so large, significant research is certainly justified.

7. Natural Gas For Transportation – The Chicken and Egg Problem

Since a significant part of this book is devoted to liquid fuels in transportation, let us consider natural gas use in that sector. Readers may have noticed natural gas powered buses and taxicabs in various cities, where its use is proudly advertised for all to see. Besides being economical, natural gas powered buses avoid that unsavory smell often associated with older diesel fueled buses.

Natural gas buses have established routes, and refueling is facilitated by the fact that buses typically return to their garages each night, where they can be refueled using special equipment. Hub-based operations are necessary because natural gas fueling stations are rare. Therein lies one problem with moving quickly to deploy natural gas vehicles, other than buses, taxis, and other special situations -- namely the lack of available refueling stations.

> *"For natural gas to be widely utilized in road transport, a refueling infrastructure would have to be built across large regions where the vehicles operate."*

For natural gas to be widely utilized in road transport, a refueling infrastructure would have to be built across large regions where the vehicles operate. That would not be a quick or inexpensive endeavor. In the U.S. there is a very extensive natural gas pipeline and distribution system in many parts of the country but not in the U.S. northeast. Will a natural gas infrastructure be added to the U.S. northeast? When? Can natural gas powered vehicles be deployed that do not operate in a large and important part of the U.S.?

This is the so-called "chicken and egg problem" fueling / vehicle sales problem. It is not possible to sell a large number of natural gas powered vehicles without the existence of a large number of refueling stations. On their own, filling station operators are unlikely to install natural gas filling pumps without there being a large number of natural gas vehicles on the road to buy their product. Clearly, the refueling infrastructure must come first,

> *"It is not possible to sell a large number of natural gas powered vehicles without the existence of a large number of refueling stations."*

which is unlikely without government mandates, subsidies, and maybe a decade for build out.

A similar chicken and egg problem was studied by the U.S. National Academies as part of its considerations of hydrogen-fuel cell vehicles. [97] A major barrier to the adoption of hydrogen is that there is no national hydrogen-fueling infrastructure to support the large-scale deployment of hydrogen-fueled vehicles. The Academies conclusion was that significant government subsidies would be required to build a network of hydrogen fueling stations ahead of the advent of hydrogen vehicles, and the cost of that endeavor would be very large.

There is potential for natural gas in fleet markets, for instance delivery trucks and vans, government vehicles, school buses, etc. Many of these vehicles operate within relatively small regions, have defined daily routes, and return to the same central location where they can be refueled.

One intriguing option is to convert school buses to natural gas fueling, which is feasible because the buses typically return to a central location, where they can be refueled. With the advent of a national liquid fuel shortage, these buses could be pressed into use for general public transportation, which would provide a ready option for the mitigation of the decline of world oil production. The problem is that most school districts could not afford the conversion and enhanced use, so state and federal governments would have to become involved.

Before a country undertakes a large national commitment to natural gas vehicles, it would be prudent -- we would say mandatory -- to establish that there is an assured long-term supply of relatively low cost natural gas available upon which to base natural gas vehicle development. However, it must be recognized that there are other options for vehicle fueling, so it is essential to objectively consider all options to determine which one or ones are likely to be the best choices.

8. Is shale gas for transportation a good idea?

Assuming that the various shale gas issues can be effectively managed, shale gas could provide an energy bonanza with profound impacts on world energy needs. How shale gas will be used could be left to the marketplace or it could be guided by government policy.

> *"Government intervention would seem to be required, if for no other reason than the chicken-and-egg problem."*

If the use of natural gas from shale is to be determined by the marketplace, then there are a number of possible outcomes. One is that its use in residential, commercial and industrial applications might expand, displacing oil, propane, and electric heating. Its use for electric power production might increase significantly, also.

Whether or not the marketplace would embrace gas for transportation on its own is doubtful. Because of the multitude of issues, government intervention would seem to be required, if for no other reason than the "chicken-and-egg problem."

However, before a government embarks on promotion of natural gas for transportation, the following questions should be seriously considered in an objective, unbiased, balanced manner:

- Is there a large, existing natural gas pipeline and distribution system upon which to build? Does the system cover the entire country or will it have to be significantly expanded?

- What are the actual total environmental impacts of large-scale shale gas development, and how do these compare to other options?

- Will a country have an adequate supply of affordable natural gas within its borders and within the borders of its immediate neighbors for the next 50-100 years?

- Will a county's natural gas resources have to be augmented by imports of natural gas by pipelines from adjacent countries or by Liquefied Natural Gas (LNG)? If so, does the related import dependence represent a national security risk? Studies indicate that the lifecycle GHG impact of LNG may exceed that of coal and other fossil fuels, which raises the question of the environmental implications of reliance on LNG. [98]

- Is the use of natural gas for selected transportation applications a better option than hybrid electric, plug–in hybrid, or straight electric vehicles?

- Is the use of natural gas for some transportation applications appropriate, such as for highway trucks, buses, and perhaps some types of fleet vehicles, for example?

- Do vehicles from one country often move into adjacent countries? If so, will those countries provide ready natural gas refueling or will vehicles have to stop at the border?

- Is the government ready and able to supply the financial support and incentives necessary to develop natural gas for transportation?

9. Post Script

As we have pointed out repeatedly, broad-scale consideration of specific energy sources and applications is required to put them into proper context before major energy decisions are made. All options have shortcomings. We often learn as we go, but careful analysis at the early stages of the development of a resource or a technology can often help avoid embarrassing discoveries later.

In the case of shale gas, many observers credit it for all of the benefits of conventional natural gas. However, the process of producing shale gas is different in that massive hydraulic fracturing is necessary in its production. A recent preliminary study out of Cornell University suggests that shale gas fracturing may emit large quantities of greenhouse gases to the surprise of many. [99] It was stated that "When the total emissions of greenhouse gases are considered, (hydrocracked, shale-sourced natural gas) and coal from mountain-top removal probably have similar releases, and in fact the natural gas may be worse in terms of consequences on global warming." For those hoping that shale gas will help with mitigating global warming, the shale gas option may deserve more careful scrutiny.

As we noted, electric power produced from natural gas is a fast-responding method of producing the electric power needed to accommodate the dramatic variations associated with the electric power created by wind and solar energy. Advocates of renewable-based electric power ought to think twice about objecting to natural gas development, because without adequate and low cost electric power from natural gas, wind and solar power may be doomed to being nothing more than bit players. More on that subject is presented in the next chapter.

"Advocates of renewable-based electric power ought to think twice about objecting to natural gas development."

C. Biofuels

1. Introduction

Liquid fuels that can substitute for oil-derived fuels can be produced from biomass. Biomass is defined as plant material such as trees, grasses, corn, wheat, soybeans, palm, sugar cane, algae, etc. The earliest biomass-based energy was burning of wood for heat and food cooking. The current interest in biomass energy is associated with the fact that it is renewable, which means new crops can be produced on a regular basis, which would keep biomass materials flowing year after year. In the past, the thought was that biomass energy had very positive environmental impacts, but recent studies have raised questions regarding that belief in some cases.

In the context of this book, we are primarily interested in biomass-to-liquid (BTL) fuels options. In the U.S., there is a large, on-going effort to grow corn and transform it into ethanol, which can be used as a gasoline substitute. In Brazil, sugar cane is grown on a large scale for the production of ethanol. Other crops are also being grown for conversion to fuels in many countries.

Before discussing some of the specific biomass-to-liquid fuel options, we begin with a few fundamentals associated with biomass plantations – crop areas that are specifically dedicated to the growth, harvesting, and processing of biomass for the production of liquid fuels.

2. Some Biomass Plantation Fundamentals

Because sunlight is diffuse, growing and harvesting large quantities of most land-based biomass requires large land areas. Harvesting biomass requires tractors and trucks that consume liquid fuels, which means that biomass-to-liquids is inherently a process that consumes liquid fuels in order to produce liquid fuels. This translates to an inherent limit on the size of land-based biomass plantations – a limit that is determined by the fuel consumption required to harvest and transport biomass that is a long distance away from facilities that can convert biomass to liquid fuels.

The most efficient geometry for a biomass-to-energy plantation is circular with a conversion facility located at the center. In this configuration, biomass is harvested and ideally transported radially to the conversion site. Why? Because a straight line is the shortest distance between two points,

> *"Because sunlight is diffuse, growing and harvesting large quantities of most land-based biomass requires large land areas."*

in this case a harvest area and the processing facility. To have a practical biomass-to-liquid fuels production system, the liquid fuels invested in system operation must be much smaller than the liquid fuels that the plantation produces. If it is not, the system is of no practical interest.

After working through the calculations for a number of biomass species, a general rule-of-thumb emerges: The radius of a biomass plantation cannot be much larger than roughly 50-miles. Thus, practical land-based biomass-to-energy plantations would be configured in a circular pattern with a 50-mile radius with conversion facilities at the center. However, the ideal is rarely possible, because there are relatively few regions in most countries that are free of obstructions, such as towns, parks, farms not part of the plantation, factories, or natural obstructions, such as hills, lakes, rivers, or rock outcrops. Furthermore, radial transport of biomass may not always be possible, because of the unsuitability of the terrain to handle heavy trucks carrying biomass to processing facilities.

> *"The radius of a biomass plantation cannot be much larger than roughly 50-miles."*

Figure XIV-1 shows the ideal circular configuration, while Figure XIV-2 shows how biomass transport might actually be implemented to avoid large obstructions, assuming that trucks do not have to run on roads. Anyone familiar with open country knows that a variety of factors can force trucks to travel on paved highways, which would further reduce the size of the plantation, because of the fuel consumption associated with longer travel to and from the outer regions. An illustration is shown in Figure XIV-3, which shows how trucks might have to travel to and from a harvest area following existing roads arranged on a square pattern, as often exists in many rural areas in the U.S. Midwest. In those situations the cost of trucking biomass will be considerably higher than the ideal, which translates to smaller biomass plantations, which produce smaller quantities of liquid fuels.

There is another important factor. A number of crops must be planted and cultivated by liquid fuel-consuming machinery before harvesting. Those functions add additional fuel requirements, which further reduce the net liquid fuels produced from a biomass plantation.

What do these and other considerations tell us about the practical size of biomass plantations? If a land-based biomass plantation is to be used for the production of liquid fuels – ethanol for instance -- then the output is limited to roughly 6-8,000 barrels per day of production. [100]

To appreciate what these numbers mean, consider the factious case of converting the total land area of a dozen U.S. Midwestern states to biomass-to-liquid production. That task would require wiping out all major cities, factory areas, and large parks, and devoting nearly all of the land for biomass-to-liquids production. This is clearly a fiction, but our purpose is to calculate illustrative numbers to provide some overall perspective.

For this calculation, we consider the total land areas of the following states: Michigan, Minnesota, Kansas, Nebraska, Oklahoma, Missouri, Wisconsin, Illinois, Iowa, Arkansas, Ohio, and Indiana. The sum total land area of these states is near 800,000 square miles, which includes lakes, rivers, and unusable areas, which we ignore for simplicity since it's a fictitious case anyway.

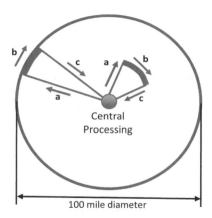

Figure XIV-1. Schematic of an ideal biomass-to-energy plantation. The outer diameter is roughly 100 miles with the central processing facility in the center. Two idealized harvesting pathways are shown. Leg "a" is the path a truck would travel out to the area to be harvested, and leg "c" is the return path for a truck filled with biomass. Area "b" is the harvested area, which is large enough to completely fill a truck with biomass.

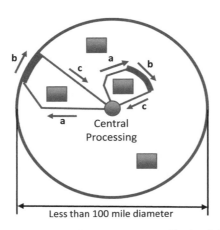

Figure XIV-2. Schematic of a more realistic biomass-to-energy plantation. The boxed areas represent towns, parks, farms not part of the plantation, factories, or natural obstructions, which block straight-line truck paths to the harvesting areas. In this illustration, truck leg "a" is lengthened by the need to steer clear of an obstruction. As in Figure XIV-1, area "b" is the active harvested area, which is large enough to completely fill a truck with biomass. An assumption in this picture is that trucks are able to follow the indicated pathways and are not required to follow existing roadways, which would further lengthen their travel distances.

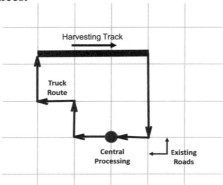

Figure XIV-3. Truck and harvesting routes in a biomass plantation in which trucks must use existing roads, which are laid out in a square pattern. In this illustration trucks follow the roads from the central processing facility to an active harvesting track, where they pick up harvested biomass and bring it to the central processing facility, after which they make another run. In this situation, the geometry differs significantly from the ideal pattern shown in Figure XIV-1.

Assuming an optimistic, steady liquid fuel production rate of 8,000 barrels per day from a 50-mile radius biomass plantation yields an approximate yield of 1 barrel per day per square mile over the course of a year. Recognize that this is a rough number. The important point is that typical liquid yields are not as low as 0.1 barrel per day or as high as 10 barrels per day.

The bottom line is that the total production from biomass-to-liquids plantations covering all 12 states would be roughly 800,000 barrels per day, which is about 4% of recent U.S. liquid fuels demand. On the basis of this fictitious, optimistic illustration, it is not remotely reasonable to believe that land-based biomass-to-liquids processes can provide significant mitigation of the decline in world oil production.

3. Corn-Ethanol: A Demonstrated Loser

Ethanol can be mixed with gasoline to form a fuel mixture that works in existing vehicles, built to operate on gasoline alone. In the U.S. ethanol is blended with gasoline at the 10% ethanol level and widely marketed. To use ethanol in higher concentrations in U.S. light duty vehicles requires vehicle redesign, because current vehicle materials would lead to accelerated engine failures.

> **"It is not remotely reasonable to believe that land-based biomass-to-liquids processes can provide significant mitigation of the decline in world oil production."**

Ethanol is the alcohol found in alcoholic beverages. It can be easily produced from sugar cane and corn by fermentation and distillation, which is the approach responsible for most of the ethanol produced in the world today. Brazil is blessed with growing conditions in which sugar cane flourishes, and that country has taken advantage of that fact to produce ethanol on a very large scale. The situation in Brazil is economically and environmentally reasonable. In addition, sugarcane-ethanol is energetically attractive, in stark contrast to corn-ethanol in the U.S. Finally, vehicles in Brazil are built to run on high concentrations of ethanol.

> *"Massive subsidies and requirements for corn ethanol in the U.S. are the result of politics, not sound energy policy. "*

> *"U.S. mandates for corn-ethanol caused the price of food to escalate."*

The U.S. produces massive amounts of corn-ethanol, but the Energy Return on Energy Investment (EROEI) for the process is poor -- some even claim that it is negative, meaning that more energy is required to produce corn-ethanol than is in the ethanol product. Massive subsidies and requirements for corn-ethanol in the U.S. are the result of politics, not sound energy policy.

The interest in corn-ethanol was rooted in the fact that it can be grown, which makes it renewable, which is a high value to environmentalists, politicians, and the public. Farmers are particularly fond of corn-ethanol because it represents a major outlet for corn production, enhanced by subsidies and mandates.

Corn-ethanol requires the use of very large areas of fertile, arable land, which would otherwise be used to grow food. U.S. mandates for corn-ethanol caused the price of food to escalate, an unwanted outcome that could have been anticipated, had there been careful analysis before the corn-ethanol mandates were imposed. At the beginning of the corn-ethanol craze, few analysts suspected that corn-ethanol would also have significant environmental problems. After all, it was renewable, which implies goodness. Today, we know that corn ethanol is environmentally marginal, which simply adds to its negatives.

Ethanol advocates describe ethanol production in billions of gallons per year. That measure yields very large numbers, which is impressive in marketing and politics. For example, the U.S. has mandated production of 35 billion gallons of ethanol per year by 2017, which is a seemingly huge number until it is translated to millions of barrels per day, which is the measure used in dealing with oil production and consumption.

Do the calculation. There are 365 days in a year and 42 gallons in a barrel, and the energy content of ethanol is about 60% of a gallon of gasoline, when properly compared. This means that 35 billion gallons of ethanol per year translates to about 1.4 million barrels per day of oil on an energy equivalent basis. Since the U.S. consumes about 20 million barrels per day of oil, that seemingly huge ethanol target boils down to about a 7% contribution in the U.S. We need what we can get, as long as it makes energy, economic and environmental sense, which corn-ethanol does not.

The corn-ethanol craze in the U.S. is the result of politics moving ahead of objective analysis. Reversing the situation will be difficult because of farm state interests, and maybe the fact that the first U.S. presidential caucus is held in farm-state Iowa every four years. No wonder so many in the real world of business fear political intrusion into energy markets.

4. Cellulosic Ethanol

Cellulose is carbohydrate plant material, which is the main constituent of cell walls in wood, grasses, cotton, and other plants. Because plants are renewable, cellulose is renewable. If ethanol or hydrocarbon fuels can be made from cellulose in an economic, energetically, and environmentally reasonable manner, cellulosic ethanol could make a useful contribution to future liquid fuel supplies.

Using a variety of feedstocks, cellulosic ethanol and other fuels could be produced in many regions of the world. Sources of cellulose are myriad and include agricultural wastes, such as corn stover and cereal straw; industrial wastes from plant materials, such as saw dust and paper pulp; and energy crops, like trees and various grasses, grown specifically for fuel production.

> *"The corn-ethanol craze in the U.S. is the result of politics moving ahead of objective analysis. "*

Note that for a number of these materials, there are existing uses, varying from fertilization of existing soils to drywall and wallboard. Removing those cellulosic materials could well cause other upsets in markets and the environment. Accordingly, related changes in use need to be evaluated. Rarely do we get something for nothing.

Growth of cellulose does not require fertile land, so it need not compete with the growth of agricultural foods, which is an important plus. The problem is that liquid fuels made from cellulose are a long way from being economic, because of the need for complex, difficult, and expensive conversion processes.

As discussed in Chapter XIII, energy return on energy invested (EROEI) and liquid fuel return on investment (LFROI) are valuable tools for assessing the potential of various types of fuels, and one of the important questions concerning the potential viability of cellulosic-based liquid fuels is whether

> *"Rarely do we get something for nothing."*

> *"It is still very much an open question as to whether cellulosic-based liquid fuels can have large enough net benefits to be of practical value. "*

the related returns are sufficiently positive. Some studies have indicated that the EROEI of cellulosic ethanol is promising, but many studies were not peer-reviewed for objectivity and accuracy, and some studies were conducted by vested interests, such as renewable and agricultural fuels organizations, environmental organizations, and units of government with potential conflicts of interest.

In reviewing studies published in peer-reviewed scientific journals, we found that results were mixed and that findings differ. Thus, it is still very much an open question as to whether cellulosic-based liquid fuels can have large enough net benefits to be of practical value. It is also likely that some countries will have greater potential than others.

5. Algae-To-Liquids

Another biomass-to-liquids concept receiving considerable attention these days is Algae-To-Liquids (ATL). In one sense, this is an attractive option, because algae make very efficient use of energy from the sun and various nutrients. However, ATL also has a fundamental liquid fuel return on investment (LFROI) problem that must be overcome.

Algae grow in water and are thoroughly soaked when recovered. A great deal of energy – usually liquid fuel energy -- is required to harvest the heavy, water-soaked algae and to remove the water. This presents a fundamental roadblock, which, to our knowledge, has yet to be overcome.

> *"A great deal of energy – usually liquid fuel energy -- is required to harvest the heavy, water-soaked algae and to remove the water."*

It might be possible to develop algae that secrete hydrocarbon liquids that are easily collected. If so, the dewatering problem could be circumvented. Since there is a good deal of venture capital investment going into algae-based liquid fuel production, there are likely other research options possible.

> *" Increasing the average mileage of passenger vehicles by 3 -5 miles per gallon would dwarf the effects of all possible biofuel production from all sources of biomass available in the U.S."*

6. Concluding Remarks

Most biomass production is currently committed to food, feed, lumber, paper pulp, fiber, etc., and biomass makes heavy use of fossil fuels for planting, tilling, harvesting and transportation of material to and from processing.

Compared with current liquid fuels use, the potential impact of successful biomass-to-liquids development would be relatively small in many locations in the world. We calculated that increasing the average mileage of passenger vehicles by 3-5 miles per gallon would dwarf the effects of all possible biofuel production from all sources of biomass available in the U.S.

Nevertheless, because all reasonable sources of liquid fuels will be needed to mitigate the impending decline in world oil production, further research and development of biomass-to-liquid options deserves priority. A number of successful biomass plantations, operating for a number of years are needed to clearly demonstrate viability. Let us not "bet the farm" until clear winners have been demonstrated and clear losers discarded.

> *" Let us not "bet the farm" until clear winners have been demonstrated and clear losers discarded."*

D. Oil Shale

1. Introduction

Oil shale is a sedimentary rock that is dark brown to black in color and high in organic matter. The organic matter is called kerogen, which is fossilized organic material that can yield liquid and gaseous fuels, when properly processed. Oil shales occur in a wide range of richness, but commercially attractive grades generally yield between 20 and 50 gallons of oil per ton of shale rock, roughly one half to one barrel per ton.

> **" Oil shale development for liquid fuels production could yield large quantities of oil products. "**

Deposits of oil shale occur around the world in such places as the United States, Australia, Sweden, Estonia, Jordan, France, Germany, Brazil, China, southern Mongolia and Russia. Estimates of global deposits are roughly 3 trillion barrels of recoverable oil, so oil shale development for liquid fuels production could yield large quantities of oil products, if processes can be developed that yield oil at reasonable prices and are environmentally acceptable.

U.S. oil shales are estimated to amount to about 2 trillion barrels of in-place oil. They are concentrated in the states of Colorado, Utah, and Wyoming, but sizable quantities also exist in the eastern U.S.

Oil can be extracted from oil shale rock by 1) mining the rock and then processing it or 2) by extracting organic material directly from the rock in-situ – in place in the ground.

To obtain useable oil, oil shale must be heated to temperatures between 400 and 500 degrees centigrade to free up useable oil and gases. Numerous approaches to surface retorting were tested at small scale during the 1970s and 1980s, and two types of surface retort facilities, vertical and horizontal, seemed to offer significant promise. One important characteristic is oil recovery efficiency and another is water requirements, which are important in semi-arid regions. No surface processing technique has yet been tested at commercial scale.

2. Oil Shale Production by Mining

Oil shales can be surface mined or deep-mined. Surface mining is used for oil shale zones that are near the surface. Surface mined shale oil would necessarily be processed in equipment on the surface. Deep mining and bringing shale rock to the surface for processing is also possible. Such an operation would resemble deep coal mining.

3. In-Situ Processing of Oil Shale

In-situ processing involves heating oil shale in-place, underground. Various approaches have been proposed and tested. True in-situ processes involve no mining: The shale is fractured, air is injected, the shale is ignited to heat the rock, and the liberated shale oil moves through fractures to production wells. So-called modified in-situ involves mining below the target shale before heating and requires fracturing the target deposit above the mined area to create void spaces for the oil and gases to escape to the surface, where they are refined.

Shell Oil Company has developed a novel in-situ heating process that shows promise for recovering oil from rich, thick shale resources lying several hundred to more than 1,000 feet deep. The process uses electric heaters placed in closely spaced vertical wells to heat the shale. Heating periods of two to four years both liquefies the kerogen and creates micro fractures in the rock that facilitate fluid flow to production wells, which bring oil and gases to the surface for processing to clean fuels and products.

Shell's slow heating is expected to improve product quality and recover shale oil at greater depths than other oil shale technologies, and the process reduces environmental impacts by avoiding subsurface combustion. An innovative "freeze wall" technology is being tested to isolate the production area from groundwater intrusion until oil shale heating, production, and post-production flushing has been completed. Shell has recently tested their concept in northwestern Colorado's Piceance Basin.

With in-situ processing, there are locations that could yield in excess of a million barrels of oil equivalent (oil and gas) per acre and require about 20 square miles to produce as much as 15 billion barrels of oil over a 40-year project lifetime.

> *" With in-situ processing, there are locations that could yield in excess of a million barrels of oil equivalent (oil and gas) per acre."*

4. Development Status

According to Wikipedia, "oil shale serves for oil production in Estonia, Brazil, and China; for power generation in Estonia, China, Israel, and Germany; for cement production in Estonia, Germany, and China; and for use in chemical industries in China, Estonia, and Russia. As of 2009, 80% of oil shale used globally is extracted in Estonia." There is currently no commercial oil shale production in the U.S.

In the 1970s there was vigorous oil shale development in the U.S., but it was interrupted in the mid 1980s by declining oil prices. Major investments by industry and government resulted in a very good understanding of oil shale resources and the development and testing of a broad spectrum of technologies for converting oil shale into liquid fuels. The lessons learned and the technologies developed from these efforts are available, as are oil shale efforts elsewhere in the world.

A critical hurdle in the development of large-scale oil shale development for liquid fuel production is the successful development and operation of first-generation plants at commercial scale. The best existing technologies for producing U.S. oil shales have not yet been tested beyond the pilot scale. Several ongoing research and pilot projects could lead to commercial scale production within the next decade. However, significant impediments and uncertainties must be resolved to attract the large private investments that will be required to advance technologies to the commercial scale. Included are access to resources on public lands, technology development and scale-up, environmental evaluation, adequate water supplies, and an accurate estimate of capital and operating costs.

Several developments occurred as a result of the U.S. Energy Policy Act of 2005, which directed the Department of Energy (DOE) to establish a commercialization program. Congress directed DOE to assess the readiness and potential of existing oil shale technologies for demonstration and implementation at commercial scale. In addition, the Department of the Interior, Bureau of Land Management has initiated a Research, Development and Demonstration Leasing Program for Oil Shale.

Several companies are currently conducting research and development that could lead to large-scale testing and commercial-scale demonstration projects within a decade under business-as-usual conditions. With the widespread realization of world oil production decline, progress could be accelerated.

5. Economics

High front-end capital investments will be required, and long lead-times will precede positive revenue streams for investors. Risks associated with the uncertainty of capital, operating costs, and environmental costs need to be reduced to facilitate capital formation and project investment.

The capital cost for an early 100,000 barrel per day surface processing facility has been estimated to be of the order of $5 billion, with later facilities likely costing less. [101] Operating costs include mining, labor, energy costs, and administration. With time, operating costs will likely decrease as operations gain experience, improved understanding, design enhancements, and

> **" High front-end capital investments will be required, and long lead-times will precede positive revenue streams for investors. "**

improved operating efficiency. Thus, economics for oil shale to liquid fuels are yet to be established, but the resource is huge, so the incentives for accelerated development will be strong, particularly when world oil production decline becomes well recognized.

E. Hydrogen

1. Overview

Hydrogen has significant potential to provide energy for various transportation applications. Many envision hydrogen as a long-term alternative to petroleum-based liquid fuels.

Hydrogen is an energy carrier, not a primary fuel such as oil, natural gas, coal, wood, etc. Hydrogen production requires investing energy from a primary fuel to separate it from some other material, typically natural gas or water. Electricity is also an energy carrier in that its creation requires a primary fuel.

> **" Hydrogen is an energy carrier, not a primary fuel such as oil, natural gas, coal, wood, etc. "**

Hydrogen can be directly combusted in internal combustion engines, like those now in use, or it can be fed to fuel cells, which directly and efficiently produce electricity. An automobile powered by hydrogen fuel cells is a type of electric vehicle in which the large battery pack is replaced by a hydrogen tank and fuel cells. Such a vehicle must be periodically refueled with hydrogen at service stations, only a few of which now exist on an experimental basis.

Fuels cells have no moving parts. They combine hydrogen and oxygen from the air to very efficiently produce electricity and their emissions are water vapor.

The challenges in going this route include 1) the need to generate hydrogen economically; 2) transporting the hydrogen to service stations, where users can refuel their vehicles; 3) building and operating hydrogen service stations; 4) improving fuel cells so they are economical and able to operate effectively in a wide range of vehicle environments; and 5) developing hydrogen storage systems that can efficiently, economically, and safely operate on vehicles.

2. A hydrogen economy will not happen easily.

As with natural gas, development of hydrogen vehicles involves a "chicken and egg problem." On the one hand, building a widespread hydrogen distribution system with service stations represents a huge, expensive undertaking. On the other hand, selling large numbers of hydrogen vehicles requires that service stations be in place for easy, convenient refueling. The distribution system must come first, but it will initially have a small number of customers, since initial sales of hydrogen fuel cell vehicles (HFCVs) will be small compared to the fleets of other vehicles on the road. The bottom line is that huge subsidies will be needed to start a HFCV program.

3. It's in the R & D phase.

Governments have long supported hydrogen-fuel cell research and development. A number of automobile companies have built prototype vehicles, and some of the major oil companies have built a few prototype hydrogen fueling stations in support of HFCV research and development.

Hydrogen is currently produced on a massive industrial scale for the refining and chemical industries. The cost of hydrogen derived from natural gas is comparable with the recent cost of gasoline. However, fuel cell development has quite a ways to go before fuel cells can provide the performance needed in vehicles at costs that consumers might be willing to pay.

> ## *"Development of hydrogen vehicles involves a chicken and egg problem."*

Safety is an issue to be addressed also. There is a big difference between an excellent hydrogen safety record achieved by well-trained professionals in industry and safety in the hands of everyday consumers. Hydrogen is flammable and explosive, so safety precautions are required at all stages of hydrogen transport, storage, and handling in consumer settings. Considerable effort and years of actual experience will be required before systems can be demonstrably considered safe.

> ## *"Safety precautions are required at all stages of hydrogen transport, storage, and handling in consumer settings."*

4. A Significant Evaluation

In 2008 The National Academies completed a comprehensive study of hydrogen fuel cell vehicles; one of us (RLH) was a committee member in that study. Highlights from the study provide a useful description of the outlook for hydrogen fuel cell vehicles: [102]

"… the maximum practical number of HFCVs (hydrogen fuel cell vehicles) that could be operating in 2020 would be approximately 2 million in a fleet of 280 million light-duty vehicles. The number of HFCVs could grow rapidly thereafter to about 25 million by 2030.

"Considerable progress is still required toward improving fuel cell costs and durability, as well as on-board hydrogen storage. The substantial financial commitments and technical progress made in recent years by the automotive industry, private entrepreneurs, and the U.S. Department of Energy (DOE) suggest that HFCVs and hydrogen production technologies could be ready for commercialization in the 2015-2020 time frame. Such vehicles are not likely to be cost-competitive until after 2020, but by 2050 HFCVs could account for more than 80 percent of new vehicles entering the fleet.

"An accelerated transition to HFCVs would require that automobile manufacturers ramp up production of fuel cell vehicles even while they cost much more than conventional vehicles, and that investments be made to build and operate hydrogen fueling stations even while the market for hydrogen is very limited. Substantial government actions and assistance would therefore be needed to support such a transition to HFCVs in the 2020 time frame, even with good technical progress on fuel cell and hydrogen production technologies. Substantial and sustained research and development (R&D) programs also are required to further reduce the costs of fuel cell vehicles and hydrogen after 2020.

"Although hydrogen could not replace much gasoline before 2025, the 25 years after that would see a dramatic decline in the use of gasoline in the light-duty vehicle fleet to about one-third of current projections, if the assumptions of the maximum practical case are met. Emissions of CO_2 will decline almost as much if hydrogen is produced with carbon capture and sequestration or from non-fossil sources.

"… alternatives such as improved fuel economy for conventional vehicles, increased penetration of hybrid vehicles, and biomass-derived fuels could deliver significantly greater reductions in U.S. oil use and CO_2 emissions than could use of HFCVs over the next two decades, but that the longer-term benefits of such approaches were likely to grow at a smaller rate thereafter, even with continued technological improvements, whereas hydrogen offers greater longer-term potential."

Thus, while hydrogen fuel cell vehicles are a potentially attractive long-term technology option for displacing gasoline and diesel fuels in transportation, the road ahead in hydrogen will be challenging and expensive. We are optimistic that the technology will succeed, but it will not happen quickly, so hydrogen fuel cell vehicles are not a practical, affordable near-term option for the mitigation of the decline of world oil production.

> *"The road ahead in hydrogen will be challenging and expensive."*

XV. Comments on Various Electric Generation Options.

A. Introduction to Electric Power

What is the most important characteristic of your electric power service? It must always be available when you need it -- on demand. Always-available electric power comes from electric power plants that operate on a variety of fuels. For the U.S. the relative contributions of the various generation types are shown in Figure XV-1.

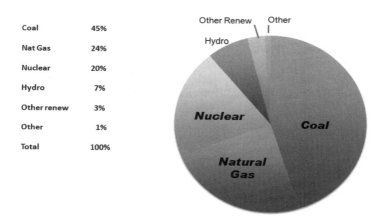

Coal	45%
Nat Gas	24%
Nuclear	20%
Hydro	7%
Other renew	3%
Other	1%
Total	100%

Figure XV-1. Net electric power generation in the U.S. from October 2008-October 2009. Roughly the contributions were coal 45%, natural gas 24%, nuclear 20%, hydroelectric 7%, and Other 4%.

In the U.S., the multitude of electric power generators are controlled by regional power authorities, which dictate how much electric power each generating plant is to provide over the course of minutes, hours, days, weeks and months. This control is facilitated by the fact that the vast majority of electric generation plants can be called on to produce electric power up to their maximum capabilities, when asked. In the jargon of the electric power industry this essential capability is called dispatchability. The system

> *"The system works because dispatchable electric power generating plants provide power-on-demand to customers very nearly 100% of the time."*

> **"Electric power is like food and water; we have to have it."**

works because dispatchable electric power generating plants provide power-on-demand to customers very nearly 100% of the time.

After the power-on-demand criterion is met, consumers want their electric power to be as inexpensive as possible. Electric power is like food and water; we have to have it, but we want it to cost as little as possible, whether we are homeowners or businesses.

Electric power production is broken down into three load categories: 1) Base (24 hours per day), 2) Intermediate (8-10 hours per day), and 3) Peak (a few hours in the middle of each day). Base and Intermediate load generators are currently fueled primarily by coal, hydro, and nuclear power with contributions by natural gas-fueled power plants. Intermediate power is provided primarily by coal and natural gas. Peak power is typically provided by natural gas, because natural gas-fired generators are easiest to ramp up and down quickly; they are the most responsive of all.

In the following, we consider and comment on the following electric power generation options: Coal, natural gas, nuclear power, wind, solar cells, and fusion.

B. Coal

1. Background

Coal is a solid hydrocarbon fossil fuel that has roughly one hydrogen atom per carbon atom, compared to roughly two hydrogen atoms per carbon in oil, and four hydrogen atoms per carbon in natural gas. As we noted, hydrogen is a very important element in chemical reactions and combustion.

Coal includes significant levels of impurities, which results in a coal ash residue and undesirable air emissions, which have been increasingly limited by regulation. For more than a century, coal has been combusted with air to produce heat for warming buildings, for various industrial processes, and for the production of electricity.

2. Resources and Production

Coal deposits suitable for mining are found in many countries, and coal has been consumed both locally and via export, as indicated in Figure XV-2, where the ten top coal producing and consuming countries are listed. In the figure energy is measured in Quadrillion Btu, a huge quantity.

Coal is mined in both surface and underground mines. Because it exists

> *"For more than a century, coal has been combusted with air to produce heat for warming buildings, for various industrial processes, and for the production of electricity."*

in a very concentrated form, coal can be mined very efficiently, and coal costs have been very low on an energy content basis, compared to the more difficult processes required to recover oil and natural gas.

The U.S. is endowed with the largest coal reserves in the world with recoverable reserves estimated to be about 270 billion tons. In 2008, the U.S. produced 1.2 billion tons of coal, second only to China. Based on EIA's 270 billion ton reserve estimate, the U.S. has more than a 200-year supply of coal at current production rates. Potential U.S. coal resources are even larger,[103] the Demonstrated Reserve Base totals almost 500 billion tons;[104] Identified Resources total about 1,700 billion tons;[105] and Total Resources approach 4,000 billion tons.[106] By any measure, the U.S. coal resource is enormous.

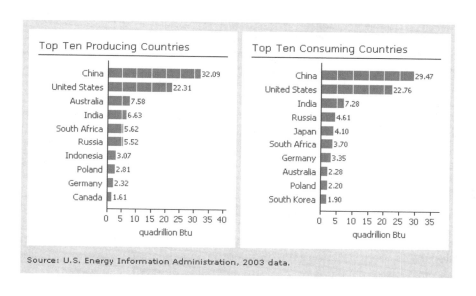

Figure XV-2. Top ten world coal producers and consumers.

3. Cleaning It Up

Coal is widely used throughout the world to produce electric power.

In earlier years, relatively modest means were used to reduce impurities in coal combustion exhausts. Over time, governments required significant decreases in coal impurity emissions, which were achieved by adding various exhaust cleanup technologies and devices. While not trivial, the costs of cleanup have often been far less than initially estimated, resulting in clean burning systems that today produce electric power at very attractive costs.

Recently, coal has been the world's fastest-growing fuel and is forecast to continue to be the primary fuel for global electric power generation for the long-term future. The International Energy Agency (IEA) notes that global demand for electricity grew nearly 25% since 2000, and it expects coal's share of the global electric power production market to increase from 6.5 billion tons in 2009 to 11 billion tons by 2030.

Various people, organizations, and governments have become increasingly concerned about the carbon dioxide emissions resulting from the combustion of all fossil fuels, because of the belief that carbon dioxide emissions contribute to global warming. Coal combustion for electric power is the highest emitter of carbon dioxide per unit of energy produced, because of coal's low hydrogen content.

Carbon dioxide emissions from the burning of coal can be reduced, but at significant cost. Two notable approaches to reduce or essentially eliminate carbon dioxide emissions from coal burning are 1) to capture carbon dioxide from the flue gas from existing coal power plants or 2) to shift to a different coal-electric technology, known as Integrated Gasification Combined Cycle (IGCC), which provides a pure stream of carbon dioxide, suitable for efficient disposal. In both cases, the captured carbon dioxide can be sequestered, which typically means that it is injected underground or underwater, where it can remain undisturbed for very long periods of time. Let us consider each of some of these options.

Separating carbon dioxide out of the exhaust gases of existing coal power plants has the advantage of continuing the operation of existing power plants, most of which are very low cost electric power generators and which often have long lifetimes. The problem is that separating carbon dioxide from exhaust gases is a very energy intensive process. The reason is that exhaust gases have high levels of nitrogen, which is a major component of the air used in combustion. Separating the majority of the carbon dioxide out of flue gas can require as much as 25-30% of the electric power from the power plant for the separation and needed compression. This technique would greatly degrade the production of electric power and increase the cost of the available power. In the extreme in the U.S., requiring rapid, wide-scale carbon dioxide separation from existing coal power plants could deprive the country of roughly 12-15% of its electric power, a dear price to pay.

A second coal-to-electric power technology option is called Integrated Gasification Combined Cycle (IGCC). In this system, coal is combusted with essentially pure oxygen, instead of air, thereby eliminating the problems associated with nitrogen. The result is a gas mixture from which nearly all non-carbon impurities can be easily removed. An added benefit is that the captured carbon dioxide that results from the processing can be relatively easily and relatively inexpensively sequestered. In many ways, IGCC is an ideal technology for taking advantage of the energy in coal, but it is more expensive than today's air combustion coal-electric power plants.

4. Sequestration of Carbon Dioxide From Coal

Sequestration involves capturing a relatively pure stream of carbon dioxide and injecting it underground into various rock formations or deep into the sea. Some of the major sequestration options are summarized below.

Preferred options involve the use of carbon dioxide for useful purposes. The most attractive of these is to pump it into older oil fields, where the carbon dioxide helps to dissolve and push some residual oil to where it can be beneficially captured; this is one form of Enhanced Oil Recovery (EOR), wherein, residual oil is produced at a modest incremental cost. In this application the carbon dioxide has a value, so it could provide financial income to an IGCC plant owner, lowering electric power costs. When almost all of the residual oil is recovered from carbon dioxide enhanced oil recovery, the oil reservoir can be sealed, containing the carbon dioxide for millions of years. After all, oil reservoirs have held oil and natural gas for millions of years, so they are excellent, demonstrated carbon dioxide storage systems.

"Sequestration involves capturing a relatively pure stream of carbon dioxide and injecting it underground into various rock formations or deep into the sea."

The problem is that the total volume of carbon dioxide that can be stored in oil fields worldwide is modest compared to the huge volumes of carbon dioxide that various groups want to sequester in the decades ahead.

Another option is to inject carbon dioxide into depleted natural gas fields. The result can be additional natural gas production, after which the gas field can be sealed, containing large volumes of carbon dioxide. Yet another option is to inject carbon dioxide into very deep, unmineable coal seams, where the carbon dioxide can displace methane that can then be beneficially recovered, after which carbon dioxide is sequestered in the coal.

Still other options include injecting carbon dioxide into underground water aquifers or into the deep ocean. These sequestration options can work but need to be studied further to ensure that containment is adequate and that there are no undesirable environmental effects.

C. Natural Gas

As we noted, natural gas is a favored fuel for new electric power generation because of its relatively low cost, the ease and speed of construction of new natural gas fired electric power generation, its modest environmental footprint, and the fact that it is the technology that is best able to effectively compensate for the intermittency of wind and solar power.

Natural gas is currently being utilized in transportation applications in a number of places. As noted, the easiest use of natural gas is in vehicles that are or can be deployed from central hubs, such as buses or taxicabs. Natural gas is used to fuel automobiles and light trucks in selected situations where special refueling options are available. For instance, our friend Tom Whipple drives a natural gas fueled Honda Civic, which he can refuel at home with his special, garage-based natural gas fueling equipment. On a trip with Tom a while back, Tom was asked what he would do if he ran out of natural gas during a trip. He said that he would call his wife to pick him up and call a tow truck to bring his car home to be refueled, because refueling elsewhere was rarely possible. Luckily, Tom has yet to face this problem, but it illustrates the importance of having widely available natural gas refueling.

> *"The easiest use of natural gas is in vehicles that are or can be deployed from central hubs, such as buses or taxicabs."*

As previously discussed, if natural gas is to be widely used as a transport fuel, government intervention would be required, if for no other reason than the "chicken-and-egg problem" of the need for established natural gas fueling infrastructure before the advent of large-scale sales of light duty natural gas vehicles or trucks.

> ## *"If natural gas is to be widely used as a transport fuel, government intervention would be required, if for no other reason than the chicken-and-egg problem."*

D. Nuclear Power

1. Introduction

Nuclear power refers to electric power plants based on the nuclear fission reaction. In today's nuclear reactors, the primary fuel is the uranium 235 isotope, which exists in nature at very low concentrations. The fission reaction occurs when a neutron strikes the heavy uranium nucleus, causing it to break into fragments, releasing large amounts of energy and roughly 2.5 neutrons, which can cause additional fissions to occur. The energy released from nuclear fissions is many millions of times the energy released in chemical reactions, such as burning coal, oil, and natural gas. Thus, much smaller volumes of nuclear fuels are needed to produce the same energy that we can obtain from huge volumes of fossil fuels.

Today's most widely deployed nuclear reactors consist of uranium fuel rods immersed in flowing water. The water serves two purposes: 1) It cools the hot fuel rods, carrying fission-generated heat to equipment that turns the heat into electric power; and 2) It provides a medium that slows down the fast neutrons that come from fission, so the neutrons can more readily induce more fission reactions. This chain reaction process is controlled by rods of material that extract neutrons from the reactor. When the rods are inserted in the reactor, they rob the reactor of neutrons, and the reactor power level declines. When the rods are withdrawn from the reactor, more neutrons are available to create more fissions, and power output is increased. During normal operation, control rod movement is modest and slow.

Besides yielding useful heat for electric power generation and neutrons for additional fission reactions, nuclear reactors produce a waste—the broken nuclear fragments resulting from fission. This waste consists of

> ## *"The energy released from a nuclear fission is many millions of times the energy released in chemical reactions."*

highly radioactive materials, which are contained in the fuel rods and must be safely stored until the radioactivity dies away in periods measured in 10-100s of thousands of years. Because each fission reaction releases very large amounts of energy per unit volume of fuel, the wastes produced are relatively small in volume. Nevertheless, the safe and proper management of nuclear waste is essential, because of its potential to damage human and animal life.

> *"The safe and proper management of nuclear waste is essential, because of its potential to damage human and animal life."*

Since higher-than-natural concentrations of uranium 235 are required in today's reactor fuel, natural uranium must be enriched to the few percent range. Further enrichment to the 90+ percent range yields material that can be used to produce nuclear weapons, which we discuss below.

Many types of nuclear reactor are possible, using uranium or other fissionable materials such as plutonium and thorium, but few of these other concepts are near commercial.

2. Current Status

Nuclear power is an important source of electrical energy in the world. Currently, there are over 430 operating nuclear power plants, producing about 17% of the world's electricity. The country with the highest share of electric power from nuclear is France at about a 75% contribution to its total electric power production.

In the U.S.104 nuclear power reactors exist in 31 states, operated by 30 different power companies. These reactors recently produced nearly 20 percent of total U.S. electrical output. In Canada, the nuclear electric contribution is about 15%. Other countries that utilize nuclear power include most of Europe, Russia, China, India, Japan, Korea, South Africa, Pakistan,

> *"Currently, there are over 430 operating nuclear power plants, producing about 17% of the world's electricity."*

and Taiwan. As of the writing of this book, hundreds of reactors are being planned and 53 are under construction. [107]

3. Outlook

The U.S. Energy Information Administration forecasts significant growth in nuclear power worldwide, while at the same time noting a number of complex issues: [108]

Electricity generation from nuclear power is projected to increase from about 2.7 trillion kilowatt-hours in 2006 to 3.8 trillion kilowatt-hours in 2030, as concerns about rising fossil fuel prices, energy security, and greenhouse gas emissions support the development of new nuclear generation capacity. High prices for fossil fuels allow nuclear power to become economically competitive with generation from coal, natural gas, and liquids despite the relatively high capital and maintenance costs associated with nuclear power plants. Moreover, higher capacity utilization rates have been reported for many existing nuclear facilities, and it is anticipated that most of the older nuclear power plants in the OECD countries and non-OECD Eurasia will be granted extensions to their operating lives.

Around the world, nuclear generation is attracting new interest as countries look to increase the diversity of their energy supplies, improve energy security, and provide a low-carbon alternative to fossil fuels. Still, there is considerable uncertainty associated with nuclear power. Issues that could slow the expansion of nuclear power in the future include plant safety, radioactive waste disposal, and concerns that weapons-grade uranium may be produced from centrifuges installed to enrich uranium for civilian nuclear power programs. Those issues continue to raise public concern in many countries and may hinder the development of new nuclear power reactors.

4. Some History

To understand nuclear power and judge its pluses and minuses, it is useful to understand some of its history.

Nuclear power blossomed as a new and exciting source of electric power following World War II. Various countries developed their own types of nuclear reactors and commercially deployed them. Some were designed with many layers of safety and accident protection, while some had relatively few layers of safety protection. Nuclear power expansion moved relatively smoothly until the first major interruption in 1979, when there was a major reactor accident at the Three Mile Island nuclear reactor complex in Pennsylvania. That accident was due to operators disabling safety systems, which resulted in a power excursion that destroyed the core of one of the reactors. Because of its in-depth safety design, the containment building did its job of containing all of the damage. Nevertheless, many in the public were frightened, and the ensuing outcry was a major setback to nuclear development in the U.S. and elsewhere.

The most severe nuclear power accident occurred in 1986 at the Chernobyl reactor in the Ukraine, then part of the Soviet Union. The Chernobyl reactor design was totally different than most other designs, and many considered it to be inherently unsafe, even before the accident. Unfortunately, there was no containment building to limit the spread of released radioactivity. The result was widespread damage to people, animals, and cities in the surrounding area. Some people died immediately and others died later, as a result of radiation poisoning. The Chernobyl accident was devastating physically and emotionally.

Since those two accidents, there have been more than two decades of safe nuclear power generation, and the world attitude towards nuclear power has gone from being generally negative to being more positive. As noted by EIA, the fact that nuclear reactor operations do not release carbon dioxide has enhanced the attractiveness of nuclear power in the minds of people and governments concerned about global warming.

As a result of the two major reactor accidents as well as instances of mismanagement by government and industry, the construction costs of nuclear plants completed during the 1980s and early 1990s in the U.S. and in most of Europe escalated. Actual nuclear power plant costs were far higher than previously forecast. Construction projects experienced long delays, which, together with increases in interest rates, resulted in high financing charges. Increased regulatory requirements also contributed to the cost increases.

More recently, a number of nuclear plants in Korea and Japan are being built on schedule and within cost, while a plant under construction in Finland, has not gone smoothly. As in all large-scale technology deployment, local circumstances can influence outcomes.

> *"Since those two accidents, there have been more than two decades of safe nuclear power generation."*

5. The Nuclear Waste Problem

As we noted, nuclear power plants produce a relatively small amount of waste, which is very dangerous and which must be properly stored for thousands of years. The world's most advanced nuclear waste program is arguably that of the French, which is the country most committed to nuclear power. Early in its nuclear power development, the French decided to use what is called a closed fuel cycle, which involves reprocessing used fuel to recover useful uranium and plutonium for re-use, in the process reducing the volume of high-level wastes that must eventually go into permanent waste storage. Recycling of reprocessed fuel allows roughly 30% more energy to be extracted from the original nuclear fuel and leads to a reduction in the volume of wastes to be stored. Recycling is expensive, and not all countries embrace its use.

France has decades of experience safely transporting spent nuclear fuel and radioactive waste by rail, trucks, and ships. The French reprocess their own spent nuclear fuel, as well as spent fuel from Belgium, Germany, the Netherlands, Switzerland, and Japan. France's longer range plan calls for deep geological disposal of high-level and long-lived radioactive wastes, and has targeted 2015 for the licensing of a permanent repository, and 2025 for its initial operation.

By comparison, the United States stores spent nuclear fuel at the reactor sites where it is produced in water tanks and robustly constructed canisters. The final storage site for spent U.S. nuclear fuel was to have been a deep underground facility in Yucca Mountain, Nevada, located in the Nevada Test Site, where nuclear weapons were routinely tested until the nuclear weapons test ban. Roughly $10 billion was spent to develop the facility. However, people and organizations objected to the facility, and in 2009 the Obama Administration proposed to eliminate all funding for the Yucca Mountain facility. In March 2010 the Yucca Mountain license application was withdrawn.

The U.S. thus finds itself without a plan for long-term spent nuclear fuel storage. While in many ways troublesome, the impasse associated with Yucca Mountain does not seem to be a near-term barrier to U.S. nuclear power development. Nevertheless, at some point the problem will require a solution.

> *"France has decades of experience safely transporting spent nuclear fuel and radioactive waste by rail, trucks, and ships."*

6. Nuclear Power Costs

Nuclear reactors in operation today produce electric power at very low prices. This is because most of these reactors are decades old, so their capital costs have been amortized to zero, which means that their power costs are based on costs for fuel, operations, and maintenance, which are collectively very low.

Recently, MIT, the University of Chicago, the Congressional Research Service, AEI/Brookings, and others performed studies of nuclear power economics. [109] Despite some differences in assumptions, these studies reached similar conclusions:

- Electric power from new nuclear plants will be more expensive than from coal or natural gas-fired power plants.
- For nuclear plants to be competitive, substantial cost reductions and federal financial incentives will be required.
- Large carbon taxes associated with global warming concerns could increase the costs of coal and natural gas plants sufficiently to make nuclear power competitive.

> *"Electric power from new nuclear plants will be more expensive than from coal or natural gas-fired power plants."*

At present, there are no firm orders for new U.S. nuclear power plants, but a number of power producers are seriously planning for new nuclear plants. In 2003, three utilities submitted applications to the Nuclear Regulatory Commission for early approval of potential reactor sites under a cost-shared program with DOE. In 2004, the U.S. Department of Energy announced cost-sharing agreements with two industry consortia to apply for Nuclear Regulatory Commission licenses to construct and operate new reactors. Since then, more than a dozen more utilities and other companies have announced plans to apply for reactor licenses, and several other companies have announced that they are considering filing applications. At the time this book was written, 17 companies and consortia in the U.S. were considering building more than 30 nuclear power plants. The Nuclear Regulatory Commission was reviewing 13 combined license applications from 12 companies and consortia for 22 nuclear power plants. [110]

The renewed interest in nuclear power has resulted primarily from the improved operation of existing reactors, and uncertainty about future restrictions on coal emissions. Nuclear power's relatively stable costs and low air emissions are attractive, particularly when combined with a substantial tax credit for nuclear generation and other incentives provided by the Energy Policy Act of 2005 and proposed legislation. New nuclear plant applications can also take advantage of amendments to the Atomic Energy Act made in the early 1990s to reduce licensing delays.

7. Nuclear Proliferation

Nuclear proliferation is shorthand for the spread of nuclear weapons (atomic bombs), fissionable materials, and nuclear weapons technology and design information. One such problem is that the spent fuel from nuclear reactors contains plutonium that can be chemically separated and used to build a nuclear bomb. A second problem is that any country that develops a capability to enrich natural uranium – increase the fraction of uranium 235 – could conceivably continue the enrichment beyond the roughly 4% needed in typical nuclear power reactors to the 80-90% needed for a nuclear bomb.

Proliferation has not been of great concern until recently, in part because countries that developed nuclear power were generally peaceful and signed the nuclear non-proliferation treaty (NPT), which has been effectively overseen by the International Atomic Energy Agency,

Outside of the NPT, India, Pakistan, and Israel have been identified as "threshold" countries in that they have developed the technology and capabilities to develop nuclear weapons. In 1998 both India and Pakistan exploded several nuclear weapons underground and both are believed to have nuclear weapon arsenals. Israel is believed to have several hundred nuclear weapons; it has neither confirmed nor denied this claim.

While North Korea signed the NPT in 1985 and agreed to the safeguards agreement with the IAEA, it later decided to abrogate its responsibilities and develop nuclear weapons, which it did. Considered a renegade state with close terrorist connections, its situation is politically extremely troubling. North Korea is an example of a country signing on and then opting out. Another example is Iran, which is widely believed to be developing nuclear weapons. How the world addresses these problems is very much an open question.

With regard to nuclear weapons technology, it is often said that the "genie is out of the bottle," which means that any country with sufficient money and talent can develop nuclear weapons. Shutting down or throttling back on civilian nuclear power development would have no impact on clandestine nuclear weapon development.

> ## *"With regard to nuclear weapons technology, it is often said that the genie is out of the bottle."*

E. Wind Energy – At least the Emperor has underwear.

1. Introduction

In the childhood story of The Emperor's Clothes, reality is recognized when a child cries out, "He isn't wearing anything at all." The "Wind Emperor" is not completely naked – There is some value to wind energy – but it comes at a very high price. The supporting fundamentals are not hard to understand, so one wonders why the realities of wind energy have not been recognized by policy-makers and the public.

Wind power is widely believed to be one of the most promising of the renewable electric generation technologies and is routinely stated to be the "most rapidly growing energy source in the world." Around the world, large government subsidies and incentivizes support the installation of wind power generation. We are concerned that so much hope and expense are being wasted, because wind energy fails in the most basic of electric power requirements,

> ## *"There is some value to wind energy – but it comes at a very high price."*

2. Wind Energy Benefits

Why the great interest in wind as a means to produce electric power? First, wind is a renewable energy source. It is not something that we are likely to deplete over time, which is not the case for many finite resource fuels, such as oil, coal, and natural gas.

Second, in operation wind generators use no water, which is in declining supply in many places in the world.

Third, there are no atmospheric pollutants emitted during the operation of wind electric generators. Of particular interest to many is the fact that wind systems do not emit carbon dioxide, which many believe is a cause of global warming. However, this seeming advantage is muted by the fact that many wind energy systems require near 100 percent fossil fuel backup in order to provide dispatchable electric power – electric power when we want it. Due to this backup requirement, the environmental impacts of wind generation can approach those of the required fossil fuel backup power sources.

3. The Realities

As we discussed, an essential feature of practical electric power is that it must always be available when we need it – It must be available on-demand. In the jargon of the electric power industry, dependable electric power generators that are available to respond to demand changes are called "dispatchable," because they are capable of dispatching electric power to where it is needed, when required to do so.

Each of us knows that winds blow at variable speeds and sometimes not at all. The physics of wind electric power generation tells us that the electric power output from a wind generator varies as the third power of the wind speed, so a factor of two difference in wind speed means a factor of eight difference in electric power generation. On this basis, small variations in wind speed lead to significant variations in electric power production. Wind changes are determined by nature, not by consumer electric power demands.

> *"Each of us knows that winds blow at variable speeds and sometimes not at all."*

The nameplate capacity of a wind generator is the maximum that it can produce, which is much higher than its average power output, because at every location the average wind speed is always much less than the maximum. The difference between nameplate power and average power output is in the 30-40% range in the best geographical regions for wind; elsewhere the average is less.

If the wind speed was constant every minute of every day and night, we would have power-on-demand. However, the wind speeds vary dramatically over the course of a year – the period over which manufacturers calculate their averages. So averages mask the periods when wind generators are not operating at all or are operating at very low levels.

> **"Averages mask the periods when wind generators are not operating at all or are operating at very low levels."**

Because of the wide variability and unpredictability of wind speeds, wind generators are not dispatchable, which means they cannot be depended on to provide power-on-demand. There are numerous examples of electric power from wind not being available when it was needed. For instance, in a "heat storm" in California in July 2006, the electric power system was severely strained due to the unusually high demand for electric power for air conditioning. At the time, the wind was blowing at low levels in the areas where wind generators were located, so available wind power was well below the rated capacity, as shown in Figure XV-3. [111] Because the local electric power grid had a high level of dispatchable electric power available, the grid did not fail, and wide area blackouts were avoided. However, had wind accounted for a larger fraction of electric power capacity, blackouts would almost certainly have occurred.

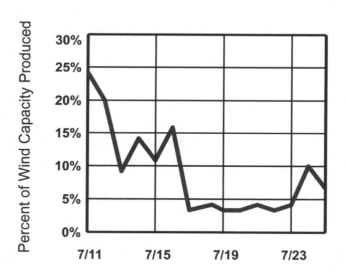

Figure XV-3. Performance of California wind power during the July 2006 "Heat Storm." When electric power was desperately needed, the wind was not blowing at a high level, and wind energy provided a small fraction of nameplate capacity for a number of days.

Low electric power production from wind has been a problem in Texas, which also has an aggressive wind power program. The regional electric power authority – The Electric Reliability Council of Texas (ERCOT) -- analyzed the capacity factor of wind and determined that only "8.7 percent of the installed wind capability can be counted on as dependable capacity during the peak demand period for the next year." In forecasts for 2009 hot summer demand periods, ERCOT estimated that 8.6% of the state's wind power capacity could actually be counted on as reliable. [112] This is a direct recognition of wind power's low availability and undependability.

ERCOT has responsibility for delivering electric power on demand, which consumers require. ERCOT's evaluation of wind's potentially reliable contributions is a tiny fraction of not only wind's nameplate capacity but also of its annualized average capacity estimate of 30-40%. This evaluation tells us that wind is nowhere near the practical, dependable power source that many believe.

The Pacific Northwest is another region with an aggressive wind program. The region's experience is that when electric power is needed most, the wind is either not blowing or only weakly available. For example, during the cold days of January 5 - 28, 2009, wind generation in the region was virtually non-existent. [113] The Bonneville Power Authority stated that over a full 56-week period, nearly a quarter of the time, total wind generation was less than 3% of total wind nameplate capacity. Unless wind is backed up almost completely with other energy sources, blackouts will occur. Can we afford to live that way?

4. Options for Rendering Wind Generation Reliable/Dispatchable

a. Introduction

If electric power is to be available on–demand, the fact that wind power is intermittent and not dispatchable means that there must always be ready backup power available for when the wind fades or dies. In other words, other electric power sources must be standing by, ready to make up for declines in wind power.

Since winds can change rapidly, the backup electric power generators must be able to respond very quickly to wind variations, if overall power-on-demand is to be maintained. Four options for providing backup are prominently discussed:

1)   Dedicated power plants close to the wind generators,
2)   The existing grid,
3)   Long transmission lines to interconnect wind generators, and
4)   Hydroelectric storage.

Let us consider each.

b. Dedicated Power Plants Close to Wind Generators

Using dedicated power plants close to wind generators would result in a wind/fossil fuel generation complex that is easily dispatchable, meaning it could be controlled by system operators much like existing power plants. In this option, the fossil plant must "run hot" so as to be able to quickly ramp up to make up for rapidly lost wind power. In utility jargon, that hot-running operation is called "spinning reserve" and it means that a small amount of fossil fuel is consumed even when the wind generators are operating at their maximum rated output.

In this option, wind acts as a fuel-saver for the fossil plant, which means that when wind power is being generated, less fossil fuel is consumed. Costs in this case are the capital costs for both the wind generators and the fossil plant plus the fossil fuel cost plus operation and maintenance costs for both generators. Obviously, power costs are much higher than claimed for the wind generators alone.

Note that the reduction in the generation of carbon dioxide in this case is not 100%, as many would like us to believe. Because wind has the capacity to deliver only 30 to 40 percent of their full power ratings in the best locations, they provide a carbon dioxide reduction of less than 30 to 40 percent, because fossil fuels are consumed to maintain the "spinning reserve."

From a practical standpoint, this configuration means that the transmission lines that connect such a wind/fossil energy electric power production complex to other regions of the electric grid system can be sized for the output of the generation complex, rather than being oversized to handle the periodic highest outputs of the wind generators.

The embarrassment here is twofold: 1) The electric generation costs of the backup power plant is less than those of the wind turbines, and 2) Since near 100 percent backup is required, the wind power generators represent

> ***"In this option, wind acts as a fuel-saver for the fossil plant, which means that when wind power is being generated, less fossil fuel is consumed."***

added cost, making the resultant electric power from the complex more expensive than it would have been without the wind generators. This is a losing proposition economically.

c. Use of the Existing Grid

The second option utilizes fossil fueled generators already on the electric grid as backup. Many existing grids have sufficient fossil fueled generation that can be kept in spinning reserve, ready to make up for the loss of power, when winds subside. With one exception, the costs of this option are similar to the first one, and wind power is acting as a fuel saver for the fossil plants. The difference is that the use of older fossil fueled generators can mean lower capital costs, when their capital costs have been amortized. Since the backbone of this method of operation is the fossil fueled generators, wind again is simply acting as a fuel saver. If fossil fuels remain inexpensive, as has always been the case, the cost of fuel saving is small, but the total cost is much higher because of the high cost of the wind generators. This is also a losing proposition economically.

d. Connection of Widely Dispersed Wind Generators with Transmission Lines

The third option involves the use of long distance electrical transmission lines to connect wind generators in widely dispersed areas of a region or country. The premise is that wind is always blowing somewhere, so that a large area interconnection can, in principle, provide constant electric power production at all times, rendering the wind network dispatchable.

For this option to be viable, thousands of miles of new, large capacity, long distance transmission lines would have to be built across a region or country. The capacity of these lines must necessarily be many times the average system power, so high levels of power can be readily moved over long distances to make up for low wind levels elsewhere.

This concept is viable in principle, but it will be very expensive because of the need for massive construction of new transmission lines that are very much oversized. Beyond the high cost of the transmission lines, there is the very sticky problem of getting the new lines permitted and built in synchronization with the construction of the large wind generation plants.

In the case of the U.S., the lines would have to cross a number of states to reach the large electric load centers on the coasts. As laws now stand, permitting these transmission lines would require the approval of states, local authorities, and impacted landowners, who have often thwarted transmission line construction in the past. To avoid related complications

> *"Beyond the high cost of the transmission lines, there is the very sticky problem of getting the new lines permitted and built in synchronization with the construction of the large wind generation plants."*

and delays, the federal government would need clear authority to mandate transmission line routes and to be able to declare eminent domain to site the lines. While the Federal Energy Regulatory Agency was recently given some related authority, it has yet to exercise it, so it is by no means clear how much time would be required to clear challenges to federal preemption. It is expected that granting this kind of federal preemption will cause substantial political conflict.

The second issue is of the "chicken and egg" variety. It is unlikely that investors will commit large sums of money to install large wind farms in areas like the northern Great Plains without the assurance that the transmission lines will be available to move the resulting electric power to markets. Clearly, it would be foolhardy to begin wind farm installation without the lines being under construction, so the "chicken and egg" problem would require federal preemption and the exercise of eminent domain ahead of wind farm construction.

The picture that emerges is of the federal government exerting unprecedented authority to enable the construction of high power electric transmission lines from coast-to-coast in conjunction with the construction of the new wind generation farms. This would require herculean efforts and federal government intrusion into state and local prerogatives. In effect, the federal government would have to clear all roadblocks and provide protections to wind investors, if such an integrated system were to result.

In the U.S. the distances to be covered by such a wind power transmission system are substantial, as illustrated in Figure XV-4. For instance,

• The distance from the northern Great Plains where much of U.S. land-based wind generation would be constructed to the Midwestern load centers is 700 – 1,000 miles.

• The distance from the northern Great Plains to the load centers in the Boston-Washington corridor is 1,200 - 1,500 miles.

• The distance from the northern Great Plains to the west coast load centers is 1,000 – 1,300 miles.

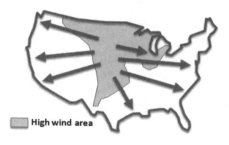

High wind area

Figure XV-4. General picture of the transmission corridors needed to move wind electric power from the U.S. Mid West to population centers.

In such an arrangement redundancy would be required to ensure that adequate transmission capacity is available during routine maintenance shutdowns, damage by extreme weather, and acts of terrorism. Redundancy will of course be costly and require additional rights-of-way for transmission lines.

It is extremely difficult to estimate how much additional transmission would be required. A proper estimate would require knowledge of where wind generators were to be installed over what time periods, where the power is being transmitted to, etc. Our estimates are that the incremental transmission required to enable such a nationwide U.S. wind energy transmission system could total 10,000 to 20,000 miles. [114] To put that estimate into perspective, the North American Electric Reliability Corporation (NERC) has estimated that by 2018 the U.S. will require about 32,000 miles of new transmission lines for expected business-as-usual operations. Thus, the transmission lines required by a large wind energy network would substantially increase U.S. transmission requirements over the next decade.

f. Hydroelectric energy storage as wind power backup

The last option we will consider involves the use of pumped hydro, which is a fancy name for large dams that produce electric power. Why pumped hydro? Because pumped hydro is the only economic method for large-scale electric energy storage.

> *"Pumped hydro is the only economic method for large-scale electric energy storage."*

Many readers are familiar with existing electric power-producing dams, where water in reservoirs behind the dams is released to flow through large water turbines, which rotate electric generators to produce electricity. Dams associated with large water reservoirs have been a backbone of electric power generation in a few places in the U.S., such as the Pacific Northwest, and the world where appropriate topographical conditions exist. If hydro sites were to be used for pumped hydroelectric power storage, the dams would have to be taken partially or completely out of service as everyday electric power generators and dedicated to wind power storage.

"Using existing dams for wind power storage would deprive existing customers of some or all of the low-cost electric power produced by the dams."

Operation would be as follows. When wind energy is abundant, part of the wind generated electric power would be used to satisfy consumer demands and the remainder would be used to pump water from below the dam to the reservoir above the dam. When wind power is inadequate to meet demands, water from the reservoir would be released to flow through the electric power generators in the dam, producing the needed electric power.

Using existing dams for wind power storage would deprive existing customers of some or all of the low-cost electric power produced by the dams. In its place would be the higher cost electric power that would be produced by wind systems.

Technologists have been searching for low-cost means of storing large quantities of electric energy for decades. A number of technologies have been conceived, and some have been tried at larger scale. All but pumped hydro has proven to be too costly to be practical. The search for viable options continues.

However, if a viable new technology were to be developed, its priority uses would not likely be to support wind energy. The reason is that inexpensive, large-scale electrical energy storage would revolutionize electric power generation. This is because it would allow all generating plants to operate at full power for much longer times than they now operate, thereby significantly lowering electric costs to consumers. In that situation, wind would end up as the high cost power generating option and would likely be even less desirable.

5. Subsidies and Other Support for Wind Power

Tax subsidies and incentives have been a critical factor in the growth of wind generation in the past, and they will likely be essential for the foreseeable future. In the U.S. in years passed, federal laws established Investment Tax Credits and Renewable Electricity Production Credits as incentives to promote certain kinds of renewable generation, such as wind. Specifically, the Production Tax Credit provided an inflation-adjusted tax credit of 1.5¢/kilowatt-hour for electric power sold from qualifying facilities, during the first 10 years of their operation.

> *"Inexpensive, large-scale electrical energy storage would revolutionize electric power generation."*

The Federal Renewable Energy Production Tax Credit currently provides a 2.1¢/kilowatt-hour incentive for the production of electricity from utility-scale wind electric power generators. It is the most important federal wind electricity incentive and has been critical in subsidizing and promoting wind generation in the U.S.

As an illustration of the importance of this credit, history shows that when the U.S. Production Tax Credit lapsed, the construction of new wind facilities dropped dramatically, as shown in Figure XV-5. Various analysts have argued that these dramatic drops in new wind generation construction demonstrate that wind energy is more of a tax play than an energy program. In any event, recent experience indicates that little wind energy would be installed without very generous government subsidies,

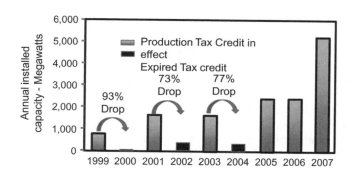

Figure XV-5. Impact of expired U.S. production tax credits on the installation of new wind capacity. [115]

Wind energy subsidies were meant to help wind energy become established commercially, after which subsidies could be removed and the wind industry would stand on its own. Wind subsidies have been in place for decades and yet when removed, wind construction dropped dramatically. This is a clear indication that wind cannot make it on a standalone basis, even after decades of improvements, and therefore, wind is inherently an uneconomic electric energy source.

To put the Renewable Energy Tax Credit into perspective, the U.S. Energy Information Administration (EIA) has forecast that the average U.S. electricity price in 2010 will be 8.3 ¢/kilowatt-hour, which means that the tax credit by itself represents an electricity production subsidy of more than 25 percent.

In the U.S., some states have renewable energy incentives that dwarf the federal incentives. For example, the state of Washington has a renewable energy feed-in tariff of 15¢/kilowatt-hour, and if the equipment is manufactured in the state, the benefit goes up to 54¢/kilowatt-hour. [116] By comparison, the current average electric rate in Washington is about 6.1¢/ kWh.

Such extreme incentives dramatically distort the electricity market, causing huge expenditures of tax dollars along with expectations and policy-making that dramatically distorts energy markets. Does this kind of extreme largesse make sense? Why would rational consumers pay taxes to promote more expensive electric power?

6. Wind Penalties

The substantial costs and complications associated with adding significant, intermittent wind electricity to the grid is beginning to be recognized. For example, in 2009 Bonneville Power Administration (BPA) in the U.S. Pacific Northwest ruled that wind generators will face a new charge over the next two years. Pending likely approval by the federal government, a new wind integration charge will be levied on all wind generators at a rate of 5.7¢/kilowatt-hour. [117] The amount of wind on BPA's system has grown rapidly in recent years, increasing both the need for backup power and the risks to system reliability. BPA has found that increased size of wind generation was compounded by the wind generators' inability to accurately account for wind changing events in their schedules, thereby requiring BPA to hold significantly larger backup power reserves in order to provide power-on-demand.

This is a significant surcharge. Current electricity rates in the Pacific Northwest are 5-6¢/kilowatt-hour, so a surcharge of 5.7 ¢/kilowatt-hour is about 100 percent.

7. The Costs of Dispatchable Wind Electricity

The costs of electric power from wind consist of the following elements:

1. The capital cost of the installed wind generators. Operating costs are typically low.
2. The capital cost of one or more backup fossil fueled power plants or the capital costs of a nationwide network of transmission lines and interconnections.
3. Fuel costs for the backup fossil fueled generation, which will be lower than if those complexes were used as standalone power plants.
4. Operating costs for the fossil fuel generation complexes or the wind power transmission system.
5. The taxpayer costs of the subsidies needed to continue the construction of future wind generation.

> *"With their tax dollars consumers are paying for wind generators that will increase their future electric bills."*

Performing reasonable cost calculations for these options depends on a myriad of assumptions and is beyond the scope of our efforts. No matter what the details, however, the cost premium for wind generated electric power is much, much higher than current consumer electric power and certainly much larger than is generally recognized.

Think about the subsidies; with their tax dollars consumers are paying for wind generators that will increase their future electric bills. Does this make sense? Does it constitute a viable long-term energy policy?

F. Solar Cells – Another Emperor's Underwear Situation

1. Introduction

Solar cells, technically called photovoltaics, are thin semiconductor devices that directly produce electricity when exposed to sunlight. As such, they turn solar energy directly into electric power without going through complicated machinery --- a distinct advantage.

The electric power output of solar cells is directly proportional to the sunlight that they are exposed to, so solar cells produce their highest electric power around noon on cloudless days, when the sun is brightest. When the sky is cloudy or it rains or snows, solar cell electric power output is dramatically reduced, typically by 60-80% for clouds and as high as 100% for snow. At night, solar cells produce no electric power, because there is no sunlight. On this basis it is easy to understand why solar cells are classified as intermittent generators of electric power. Like wind energy, solar cell power is not dispatchable and cannot provide power-on-demand without electric energy storage or a backup power plant.

Some personal observations by Bob Hirsch:

> *When I was a teenager, I bought my first solar cell -- an exposure meter to use in conjunction with my camera. Today the two are packaged as a single unit, but that was not the case many decades ago. I felt then, as I do today, that solar cells are a truly marvelous technology!*

> *Later in life, I had the privilege of managing the federal solar energy program in the U.S. Energy Research and Development Administration, the predecessor of the Department of Energy. In that capacity, I was given a small solar cell device packaged with a power meter to show how much electric power was being generated, when exposed to sunlight. Over the ensuing years, I periodically used my little device to measure solar cell electric power generation in different places at various hours during the day, so I have a first-hand feeling for how solar cells operate under different conditions. I remain enamored with the technology, but that is not the same as pragmatically evaluating how it fits into the commercial electric power world.*

2. Solar Cell Positives

Because solar cells are direct conversion devices (sunlight in and electric power out), complicated machinery to generate electric power is not needed, as we mentioned. Typically, the primary maintenance is to periodically wipe off any dust, dirt, or snow that accumulates on the cells, reducing power output.

Typically, solar cells arrays do not require cooling water. In a few applications, mirrors are used like a magnifying glass to concentrate sunlight. In so doing, inexpensive mirrors are substituted for expensive solar cells. Cooling by water or other means is needed in those situations.

One very attractive application of solar cells is in the production of electric power in remote locations, where other forms of generation are difficult to install, refuel, and maintain. In those situations, a large battery can be packaged with some solar cells, providing power on demand in many

> *"One very attractive application of solar cells is in the production of electric power in remote locations, where other forms of generation are difficult to install, refuel, and maintain."*

situations. While expensive, the combination provides an electric power generator that requires very little maintenance and does not need fuel to be trucked in over what are often long distances. This combination can be a boon in rural areas and in underdeveloped countries.

Another positive feature of solar cells is that they emit no pollutants or carbon dioxide during operation. However, this advantage is blunted by the fact that in large-scale applications, solar cells require 100 percent fossil fuel backup in order to provide the dependable electric power that the public demands.

3. Solar Cell Negatives

By far the biggest drawback to the widespread use of solar cells is the fact that they cannot produce electric power on demand; they are intermittent electric power generators because sunlight is intermittent.

Another disadvantage of solar cells is that they are expensive. Over the decades, research and development have brought prices down considerably. With sales of solar cells increasing, primarily due to subsidies, manufacturing facilities have been expanded, allowing for the cost savings that comes with higher volume production.

In most applications, solar cells must last a very long time to make financial sense. A lifetime target of 20 years or more without significant degradation is often considered necessary. Solar cell lifetimes can be diminished by sunlight, which can damage the delicate materials that solar cells are made of. It is difficult to run accelerated life tests on solar cells, so questions about lifetime exist, particularly when new solar cell materials and new fabrication techniques are utilized. In addition, weather conditions in various locations can cause cell surface or bulk degeneration. For example, sand storms can damage surfaces and connections, and dramatic temperature swings can also be problematic.

4. Options for Rendering Solar Cells Reliable and Dispatchable

The fact that solar cell power is intermittent and not available-on-demand means that there must always be ready backup power available for when sunlight is less than maximum, during the day and at night. In other words, electric energy storage or electric power sources must be standing by, ready to make up for declines in solar cell power on short notice. As discussed in the wind section, reasonably priced, large-scale, practical electric power storage is not widely commercially available, so backup electric power plants must be ready to quickly make up for lost solar cell power.

Effective solar cell backup can come from the existing electric grid or dedicated power plants. Using the electric grid as backup is in effect a tax on all other electric power users, because solar cell power is poaching on the electric grid that already exists in most places and is being paid for by other consumers. Grid backup works as long as solar cell power is a small fraction of the power handled by the electric grid. The situation with solar cells is similar to what we discussed in the wind section.

For large-scale power, solar cells must be tightly coupled to reliable standby electric power generators capable of quickly ramping up, when solar cell power is degraded by clouds, rain, snow, and night. That kind of electric power generation must be fossil fuel based, because only fossil fuel electric generation is capable of the required rapid response. As with wind, solar cells simply act as a fuel saver for the fossil fuel backup. If fossil fuels remain inexpensive, as has been the case, the cost saving is small, but the overall cost to the consumer is high.

5. Subsidies and Other Support for Solar Cells

As we discussed in the wind section, tax subsidies and incentives have been an important factor in the growth of solar cell generation in the past, and they will be essential for the foreseeable future.

Solar cell energy subsidies were initially meant to help the technology become established commercially, after which subsidies could be removed and the solar cell industry could stand on its own. That is not likely to happen in our lifetimes, because solar cells are not standalone viable for large-scale electric power production. The technology is inherently uneconomic except in niche applications.

6. The Costs of Dispatchable Solar Cell Electricity

The cost issues associated with solar cells parallel those of wind energy:

- The capital cost of the installed solar cells. Operating costs are generally low.
- The capital cost of backup fossil fueled power plants.

> ## "The cost premium for solar cell generated electric power is much, much higher than today's electric power costs."

- Fuel and operating costs for the fossil fueled electric power generation required for 100% backup, required at night.
- The cost of the subsidies necessary to continue the construction of future solar cell generation.

Reasonable cost calculations for these options depend on a myriad of assumptions and are beyond the scope of our efforts. No matter what the details, however, the cost premium for solar cell generated electric power is much, much higher than today's electric power costs. As noted previously, consumers are contributing their tax dollars for solar cell subsidies that will increase their future electric bills, if and when solar cell power becomes a significant part of the power grid. Again, we ask whether this arrangement makes sense? We think not.

G. Fusion Power – Still Waiting and Waiting

1. Introduction

Nuclear fusion reactions provide the energy in the sun, the stars, and the hydrogen bomb. Fusion fuels include isotopes (varieties) of the light elements hydrogen, helium, lithium and boron, which are abundant and inexpensive. Fusion power is in the research stage and has been since the 1950s. The ultimate goal is to produce power plants that can produce electric power that is clean and essentially inexhaustible, because the fuels are essentially inexhaustible.

To produce fusion, gases of fusion fuels must be heated to hundreds of millions of degrees and held long enough for much more energy to be released by nuclear fusion than was invested to do the heating. At these extremely high temperatures, fusion gases are so energetic that fuel atoms are torn apart into a mixture of electrons and nuclei, which is called a gaseous plasma. Neon signs are an example of cool plasmas.

Containment of fusion fuels on the sun is by gravity. Since gravity is not usable for fusion on earth, researchers have used magnetic fields, electrostatic fields, and inertia to provide containment. Thus far, no magnetic or electrostatic fusion concept has demonstrated success. Only inertial fusion in the form of the hydrogen bomb has proven practical in a weapon that we hope is never used. In the following, we touch on the three actively studied containment techniques: Magnetic, Inertial, and Electrostatic.

2. Mainline Magnetic Fusion

In magnetic confinement fusion research, the world fusion physics community has converged on a toroidal (donut-shaped) "magnetic bottle" known as the "Tokamak," a Russian pioneered concept. The magnetic fields in principle can contain the hot fusion plasma. Increasingly larger tokamak experiments have been built over the years, and the focus is now on a very large new tokamak experiment, called ITER, which stands for International Thermonuclear Experimental Reactor. It will be built in France by a consortium of countries. . If successful, ITER will burn isotopes of hydrogen and produce more energy than required to create and sustain its fusion plasma. [118]

One of us (RLH) ran the U.S. government fusion program in the early to mid 1970s and helped import the tokamak concept into the U.S., because it seemed like a promising concept at the time. Analysis in later years showed that tokamaks would not make viable commercial power systems, but the concerns went unheeded. When dedicated proponents take over a government research program, it can be very hard to change course, even though the facts dictate.

> *"When dedicated proponents take over a government research program, it can be very hard to change course, even though the facts dictate."*

Today, the outlook for success with the physics aspects of ITER is good, but the likelihood of commercial success is near zero. Thus, when ITER operates, "the operation might well be a success but the patient will in effect be dead." As a result, the world will have wasted decades and tens of billions of dollars on a dead-ended concept. Sadly, the ITER waste could have been avoided.

3. Inertial Fusion

In the hydrogen bomb, containment is "inertial." What that means is that the rapid compression of fuels to extremely high densities and temperatures results in an extremely large fusion energy release before the materials fly apart. While inertia holds the materials for just a tiny fraction of a second, it can be long enough for much more energy to be released than was invested in the compression.

The major challenge for inertial fusion has been the need to build and operate an extremely large laser to compress and ignite small fusion pellets in the laboratory. After the construction and operation of a succession of ever-larger lasers, a super-giant laser has recently been completed; it's called NIF for National Ignition Facility. [119]

> *"There are large engineering problems that must be overcome if a successful physics experiment is to be transformed into a commercial electric power source."*

Serious experiments with this huge machine may well have achieved large fusion energy releases by the time that this book is published. If so, that achievement will be of great scientific significance and will likely be heralded in the media as the gateway to endless fusion energy.

However, there are large engineering problems that must be overcome if a successful physics experiment is to be transformed into a commercial electric power source. First and foremost, a huge, rapid fire, dependable laser will have to be developed to provide the needed bursts of light energy into a chamber where fusion pellets can be made to explode at a high rate. That and other challenges will be very difficult and will take time. How long that might be, how much it might cost, and how commercially robust the final product might be are not known at this time. We are guardedly optimistic and hopeful for this approach.

4. Electrostatic Fusion

This concept was conceived by Philo T. Farnsworth, the inventor of television. Pioneering work was performed in the 1950s and 1960s, and the approach was proposed for federal government funding in 1967. The proposal was turned down by a committee of fusion laboratory physicists, who had legitimate scientific questions but were concerned about potential competition from ideas others than their own. While the future of the concept is uncertain, it offers the potential for a small power system, suitable for commercial electric power and compact power for the military or for space exploration. [120]

5. Conclusions

Serious fusion research has been going on for over half a century, and yet significant scientific challenges still remain. Scientific success in the laboratory will be a major achievement but must be followed by difficult development and engineering before practical success can be claimed. We have great hope for commercial fusion power but have no basis for believing that it will happen any time soon.

H. Other Technologies

Over the course of the decades a number of other electric power technologies have been proposed. Some have been studied and some advanced to an early experimental stage. New technologies and innovations justify reconsideration of older concepts. In some cases, we observe that inexperienced researchers seem to be ignoring the lessons learned in the past.

Some of the technologies recently discussed are the Space Power Satellite (SPS), which would put huge fields of solar cells into orbit above the earth. Previously, it was concluded that related costs would be extremely high and beaming the energy back to earth would be dangerous.

Ocean wave power periodically emerges but has thus far been deemed too expensive. Ocean Thermal Energy Conversion (OTEC) would take advantage of the small temperature difference between surface and deep waters. OTEC was found to be too expensive, subject to fouling by ocean organisms, and environmentally questionable. And the list goes on.

We strongly support periodic reevaluations of these and other technologies, because new science and new perspectives can sometimes change previous evaluations. Keep in mind that new ideas are often very exciting and appealing. However, new ideas in energy must be brought through the research, development, and demonstration stages before they can be deployed. And it is deployment that provides the energy that we consume.

"New technologies and innovations justify reconsideration of older concepts."

XVI. Winners and Losers When World Liquid Fuels Production Declines

A. Introduction

When world oil production goes into decline, exporting countries are certain to be better off than importing countries. In order to provide some insights into how world events might unfold after the decline of world liquid fuels production, let us first consider the situation for the world's largest consuming countries, the largest importers, and largest exporters.

What follows are data that we have rounded off, because our interest is in extracting insights into what is likely to happen a few years hence. Furthermore, available data are not always the most recent or accurate, and rough numbers suffice for our purposes. Finally, some of the export/import numbers are complicated by the fact that some countries export crude petroleum and import finished products (gasoline, diesel fuel, lubricants, etc.), because they do not have the refining capacity to satisfy their internal finished product needs. For example, Iran is a major oil exporting country but needs to import gasoline, because it cannot produce enough with its existing refineries.

> *"When world oil production goes into decline, exporting countries are certain to be better off than importing countries."*

B. Largest Liquid Fuel Consuming Countries

As we have noted, recent world liquid fuels consumption was roughly 85 Million barrels/day (MMbpd). Rather than provide details for over 200 countries, we focus on the largest liquid fuels consumers, shown in Table XVI-1, where their recent annual consumption, annual production, and annual imports are provided. Two of the top consumers on the list are exporters, namely Russia and Mexico. Of the largest consumers, a number have very little or no liquid fuels production within their borders, e.g., Japan, Germany, South Korea, France and Spain.

C. Largest Liquid Fuel Producers

As shown in Table XVI-2, the major oil producers are Saudi Arabia and Russia with the United States a close third. However, because of its high consumption rate, the U.S. ends up as the world's largest importer. The fifth largest liquid fuels producer is China, which is also a significant importer and whose imports are forecasted to grow rapidly in the future. We will have more to say about China shortly.

There are a number of data sets available providing estimates of exports, imports, and consumption; we have chosen a few for our purposes here. [121]

Table XVI-1. Largest liquid fuels consuming countries, their recent production, and their recent import dependence. *Million barrels day (MMbpd)*

Country	Consumption	Production	Imports
United States	19.4	6.7	11.3
China	8.0	3.8	3.6
Japan	4.8	-	5.0
India	2.9	0.8	2.5
Russia	2.8	9.9	Net Exporter
Germany	2.5	-	2.9
Brazil	2.4	1.9	0.4
Canada	2.3	3.2	Net Exporter
South Korea	2.3	-	3.0
Saudi Arabia	2.2	10.8	Net Exporter
Mexico	2.1	3.2	Net Exporter
France	1.9	-	2.4
United Kingdom	1.7	1.5	1.5
Italy	1.7	0.1	1.9

> **"The world's largest oil producers are Saudi Arabia and Russia with the United States a close third. However, because of its high consumption rate, the U.S. ends up as the world's largest importer."**

Table XVI-2. Largest liquid fuels producing countries, their recent consumption, and their recent exports *Million barrels/day (MMbpd)*

Country	Production	Consumption	Exports
Saudi Arabia	10.8	2.2	9.8
Russia	9.9	2.8	5.1
United States	6.7	19.4	Net Importer
Iran	4.3	1.7	2.6
China	3.8	8.0	Net Importer
Canada	3.2	2.3	2.0
Mexico	3.2	2.0	1.7
United Arab Emirates	3.0	0.5	2.2
Kuwait	2.8	0.3	1.7
Venezuela	2.6	0.7	1.4
Norway	2.5	0.2	2.0
Brazil	1.9	2.4	0.4

D. Some Special Situations

1. China

All liquid fuel importing countries will be negatively impacted by the impending decline in world liquid fuels production; they will endure growing shortages, as world supplies shrink. The hardest hit will be the U.S., China, Japan, India, South Korea, most countries in Europe, and many smaller countries.

> *"The hardest hit will be the U.S., China, Japan, India, South Korea, most countries in Europe, and many smaller countries."*

China deserves special note in this context, because it has been on a spending spree in recent years to lock up oil production in a number of countries around the world. It should be noted that China is also investing to ensure imports of a number of strategic materials from other countries.

Resource rich countries have abrogated contracts with private companies in the past. For example, decades ago many OPEC countries nationalized their oil fields, kicking out the International Oil Companies (IOCs), which had valid contract concessions. Recently, Venezuela in effect did the same thing with a number of IOCs. Will resource-rich countries do the same to Chinese companies in the future? We think that is unlikely, because of the fundamental difference between an ExxonMobil and a commercial arm of a major world economic and military power. Kicking ExxonMobil out of a country incurs little if any financial loss to the host country. Kicking a Chinese government entity out of a country is to incur the wrath of a major military power, risking the possibility of Chinese military action or other kinds of strong interventions. It is not inconceivable but unlikely.

We can conceive of three likely reasons why China has been investing so heavily in oil production in other countries:

1) China tends to do serious national long-range planning and invest where appropriate;

2) China recognized the impending decline in world liquid fuels production at least 4-5 years ago and recognized that investment in liquid fuels production in other countries was very much in their national

security interest;

3) China has accumulated large amounts of dollars from many years of export surpluses with the U.S. China recognizes growing U.S. financial problems, due to profligate spending and growing debt. These actions are likely to lead to a falling value of the U.S. dollar in the not-too-distant future, so investing its accumulated dollars before they decline in value makes very good sense for China.

2. Oil Exporters

From the point of view of oil exporters, the decline of world liquid fuels production means a large increase in oil prices, which means huge new financial windfalls. Those developments cut both ways:

- Oil exporter wealth will increase dramatically. The wise investment of that wealth could greatly strengthen their long-term security and the well being of their citizens and governments. Norway has managed its oil wealth wisely for example. [122]

- Growing world oil shortages with attendant high oil prices could easily lead to wars aimed at capturing oil production from exporting counties. Indeed, history is replete with examples of resource wars.

How this all plays out is an open question.

XVII. Global Warming – What a Mess!

A. Introduction

We do not know whether global warming has been or will be caused by man-made carbon dioxide emissions. However, after many years of following the literature on the subject, we are concerned by a number of troubling issues: 1) The absence of a temperature rise over the last decade; 2) Inconsistencies and inaccuracies regarding various earthly changes; 3) Potential research data manipulation and contradictions; 4) The politicization of the science; and 5) Some unprofessional and ethically questionable behaviors by both global warming analysts and special interest groups.

> *"We do not know whether global warming has been or will be caused by man-made carbon dioxide emissions."*

The proposition that climate change is produced by human activity is called the Anthropogenic Global Warming Theory (AGWT). In the following, we point out a number of our concerns. Our purpose is to provide you with some food for thought, not to reach a definitive conclusion on an extremely complex and highly controversial issue.

At the outset, let us say that fairly unambiguous data indicate that global warming occurred during the latter half of the last century. The big question is how much of recent warming was due to human activities. Proponents of global warming mitigation insist that the problem is human activities and that we need to take immediate action to avert dangerous warming in future years. Skeptics point out that geologists have determined that the earth has naturally warmed and cooled for millions of years before humans became a potential cause of any changes. Skeptics also contend that the science that forecasts a catastrophically warm future is both incomplete and flawed in many instances.

Proponents of AGWT urge nations to dramatically reduce their carbon dioxide emissions to avert future climate disaster. Objective analysts tell us that it would be enormously expensive to undertake the proposed changes.

> *"The big question is how much of recent warming was due to human activities."*

We believe the costs would result in major damage to economic development and that extreme caution is advised prior to incurring such costs.

Another problem is that all of the world's major carbon dioxide emitters must agree to work in concert for total reductions to be sufficient to hold down carbon dioxide concentrations in the atmosphere. Countries would have to divert a significant amount of their national financial resources to pay for the changes. Although there have been efforts to arrive at international agreements, many of the world's major carbon dioxide emitters refuse to take the drastic actions that have been proposed. We believe that countries that might decide to go it alone in reducing their carbon dioxide emissions could severely disadvantage their economies and their position in international trade.

Our concerns and issues are as follows.

B. Some Things That Don't Jibe

1. Global Temperatures

Verification is crucial in scientific research and forecasts, but the verification of global warming in recent years is problematic. Temperatures have been flat or slightly declining over the past decade. [123] That was not supposed to happen because carbon dioxide concentrations in the atmosphere have increased over that period. If the relationship between CO_2 and temperature is as direct we have been told, then there should have been a noticeable global temperature increase.

> *"Temperatures have been flat or slightly declining over the past decade."*

After the fact, some climate change researchers now contend that the current situation is within the bounds of model expectations. However, the Intergovernmental Panel on Climate Change (IPCC) 2007 report told us that without major warming by 2010, the results will fall outside recent temperatures. [124] As an aside, if temperatures had increased instead of remaining flat over the past decade, we suspect that global warming proponents would have contended that the warming would have been added proof that their global warming theory was correct. [125]

2. Hurricanes

One rationale for addressing global warming is to slow increases in tropical hurricane intensities; Hurricane Katrina is sometimes cited as an example of what global warming portends. However, Katrina made landfall as a category three storm (five being strongest), and did most of its damage to a city built below sea level with faulty levees. Stronger storms have struck North America before; it has been estimated that the most destructive hurricane to strike the U.S. was the Miami hurricane of 1927, well before global warming was claimed to have started in earnest. [126] Since the 1990s, overall hurricane activity has been decreasing, which contradicts what some global warming proponents expected.

3. Ice Caps

In 2007, the Northern Hemisphere reached a record low in ice coverage and the Northwest Passage reopened. Global warming proponents stated that melting was occurring faster than expected and that aggressive efforts to combat warming were needed. However, ice coverage data are only available back to the late 1970s. Prior to that, the satellite technology to measure a real ice extent did not exist, but it is known that the Northwest Passage has been open before.

We had been told that a cooling in Antarctica was consistent with global climate models until recent studies found that the opposite was true. These inconsistencies do not spur confidence.

4. The Sun

The second half of the twentieth century, which experienced warming, was during a major solar activity maximum period, which is now ending. Total solar irradiance has been steady or declining in parallel with global temperatures over the past decade. A number of analysts have argued that solar activity correlates well with observed earth temperatures, but no mechanism has been advanced and accepted to prove causation. Nevertheless, the correlation between solar activity and earth temperatures is remarkably strong. It is a troubling open question.

5. Unusual Weather

In the popular press, the question is often posed "But what about all this crazy weather we've been having lately?" People seem to forget that there

> *"A number of analysts have argued that solar activity correlates well with observed earth temperatures, but no mechanism has been advanced and accepted to prove causation."*

have often been remarkable weather events in the past that have not been rivaled recently.[127] Whether it was the 1900 Galveston Hurricane, the 1889 Johnstown Flood, or even the worst tornado outbreak in history in 1974, unusual weather events occurred long before the recent carbon dioxide buildups.

6. That Scientific Consensus

We are repeatedly told that there is virtually 100% agreement among scientists that global warming is due to human activities. Well, that is not quite true.

For example, starting in 1999, a petition was circulated objecting to the theory of man-made warming and to the proposed Kyoto Climate Treaty. Credentialed scientists and engineers were sought to support what became known as the Oregon Petition Project, aimed at expressing an opposing viewpoint. The organizers required potential signatories to fill out a form, stating their college-based technical degrees, signing the form, and mailing it in.

> *"The number of signatory skeptics was estimated to be twelve times larger than the number of scientific reviewers claimed by the UN Intergovernmental Panel on Climate Change."*

Over 31,000 technically trained skeptics of the theory of man-made global warming signed the petition. More than 9,000 held technical PhDs. The number of signatory skeptics was estimated to be twelve times larger than the number of scientific reviewers claimed by the UN Intergovernmental Panel on Climate Change. [128]

The petition read as follows:

We urge the United States government to reject the global warming agreement that was written in Kyoto, Japan in December 1997, and any other similar proposals. The proposed limits on greenhouse gases would harm the environment, hinder the advance of science and technology, and damage the health and welfare of mankind.

There is no convincing scientific evidence that human release of carbon dioxide, methane, or other greenhouse gasses is causing or will, in the foreseeable future, cause catastrophic heating of the Earth's atmosphere and disruption of the Earth's climate. Moreover, there is substantial scientific evidence that increases in atmospheric carbon dioxide produce many beneficial effects upon the natural plant and animal environments of the Earth.

While signatory names were screened to verify validity, some bogus names made the list, believed submitted by those wishing to discredit the petition. A number of blogs and articles did their best to discredit the petition, but the language in the ones we reviewed were emotional, rather than scientific in tone.

The petition cover letter was prepared by Dr. Frederick Seitz, Past President, U.S. National Academy of Sciences, President Emeritus of Rockefeller University, and Nobel Prize winner in physics – not a scientific slouch.

> *"There were a number of interchanges that showed data manipulation and outright dishonesty among some significant researchers."*

Thus, the next time you see a statement regarding the so-called scientific consensus about human causation, you might question the knowledge and objectivity of the source.

C. Climategate

In virtually all scientific research fields, there is currently close communication between researchers via the Internet. In the case of climate change/global warming research, Internet communication was a boon, until it became an embarrassment.

In November 2009, an unidentified person or organization hacked into the climate change email archives of the University of East Anglia in the UK. The university was a major center for research and coordination for world climate change research. Over 1,000 private emails sent to and from the university's Climatic Research Unit (CRU) were posted on the Internet, along with a multitude of other documents. In addition to the normal give and take, which is characteristic of good research, there were a number of interchanges that showed data manipulation and outright dishonesty among some significant researchers. [129]

The university's climate research unit director, Professor Phil Jones, was accused of manipulating data and withholding scientific information to prevent its disclosure. He subsequently relinquished his post pending the completion of an investigation.

Among others about whom questions were raised was Professor

Michael Mann, a prominent climate researcher at Pennsylvania State University. A special panel of inquiry was appointed by the university to look into allegations of misconduct on his part.

Roughly a decade ago, Professor Mann produced an analysis of past climate data purporting to show that after a thousand years of temperature decline, the earth's temperature was suddenly increasing dramatically. This so-called "hockey stick" was both dramatic and frightening. It was given prominent attention and featured by the IPCC.

Subsequent events were described as follows: [130]

> *Since 2003, ... when the statistical methods used to create the "hockey stick" were first exposed as fundamentally flawed by an expert Canadian statistician Steve McIntyre, an increasingly heated battle has been raging between Mann's supporters, calling themselves "the Hockey Team", and McIntyre and his own allies, as they have ever more devastatingly called into question the entire statistical basis on which the IPCC and CRU (Climate Research Unit) construct their case.*

> *Back in 2006, when the eminent US statistician Professor Edward Wegman produced an expert report for the US Congress vindicating Steve McIntyre's demolition of the "hockey stick", he excoriated the way in which this same "tightly knit group" of academics seemed only too keen to collaborate with each other and to "peer review" each other's papers in order to dominate the findings of those IPCC reports on which much of the future of the US and world economy may hang. In light of the latest revelations, it now seems even more evident that these men have been failing to uphold those principles, which lie at the heart of genuine scientific enquiry and debate.*

We do not know who, if anyone, was guilty of what. However, it is clear from reading the email texts that what was going on was not a demonstration of high scientific integrity and unbiased research.

In addition to the embarrassment of the hacked emails, some bogus claims in the 2007 Intergovernmental Panel on Climate Change (IPCC) report came to light. For instance, [131]

> *The IPCC claims that the Himalayan glaciers could melt away as soon as 2035. The forecast was based on a media*

> *"What was going on was not a demonstration of high scientific integrity and unbiased research."*

interview with a single Indian glaciologist in 1999, and the Indian glaciologist who was interviewed, Syed Hasnain, says that he was misquoted; indeed he had provided no date. *Professor Hasnain discovered the mistake in 2008 when he read the IPCC's published report, but he said: "There are many mistakes in it. It is a very poorly made report. . . . My job is not to point out mistakes. And you know the might of the IPCC. What about all the other glaciologists around the world who did not speak out?"*

The IPCC warned that because of global warming, the world had "suffered rapidly rising costs due to extreme weather-related events since the 1970s." They cited one study to support their claim, but when the research was published in 2008, after the IPCC report was released, the study noted: "We find insufficient evidence to claim a statistical relationship between global temperature increase and catastrophe losses."

The IPCC warned that up to 40 percent of the Amazon rain forest might be wiped out by global warming, but the sole source for that claim was a non-refereed report authored by two people who the Sunday Times of London referred to as "two green activists," one of them with the World Wildlife Fund.

The IPCC even got wrong the percentage of the Netherlands that is below sea level. The report claims that the percent is 55 percent, when the right number is 26 percent.

Other criticisms abound. For example: [132]

A Canadian analyst has identified more than 20 passages in the IPCC's report, which cite similarly non-peer-reviewed WWF or Greenpeace reports as their authority, and other researchers have been uncovering a host of similarly dubious claims and attributions all through the report. These range from groundless allegations about the increased frequency of "extreme weather events" such as hurricanes, droughts and heat waves, to a headline claim that global warming would put billions of people at the mercy of water shortages - when the study cited as its authority indicated exactly the opposite, that rising temperatures could increase the supply of water.

> ## "For the past 15 years there has been no 'statistically significant' warming."

Finally, shortly after stepping down from his leadership post at the University of East Anglia as a result of the Climategate debacle, we learned, "Professor Jones ... conceded the possibility that the world was warmer in medieval times than now – suggesting global warming may not be a man-made phenomenon. And he said that for the past 15 years there has been no 'statistically significant' warming." [133]

Without careful analysis by an unbiased independent body, the validity of the various allegations is open to question. Perhaps the IPCC made some innocent errors, but a number of the mistakes clearly raise credibility questions. Indeed, the errors that we have seen are all in the direction of making climate change forecasts more dire. A strong bias was evident.

D. Intolerance of Dissent

AGWT advocates often contend that we should not question the "settled science" due to the moral imperative of "saving the planet". Further, the AGWT establishment has attempted to silence critics by withholding funding and preventing their publication in peer-reviewed journals. This behavior is troubling and seems indicative of religious fervor, rather than honest scientific debate. Those questioning the AGW theories have been derided as being the equivalent of "holocaust deniers," believers in a flat earth, and believers that NASA faked the moon landings. The famed economist Paul Krugman accused deniers of being guilty of "treason against the planet." [134] Some supposedly serious scientists and environmentalists have suggested that AGWT skeptics be fined, jailed, or put on trial for "crimes against humanity and nature." [135]

If the science is indeed as unassailable as many contend, then why are they so sensitive to criticism and debate?

E. Geoengineering Options

There are options for modifying the climate other than the massive worldwide reduction of carbon dioxide emissions. A number of these options are collectively called "geoengineering," which the U.S. National Academies defines as "options that would involve large-scale engineering of our environment in order to combat or counteract the effects of changes in atmospheric chemistry." [136] Some of these technologies offer the opportunity to modify the climate at a much lower cost and much faster than carbon dioxide reductions. Each option has its strengths and weaknesses, and each has environmental effects that must be carefully considered, lest solving one problem might create others. The fact that many environmentalists avoid, even exclude consideration of geoengineering options is troubling and raises questions as to their interest in attacking climate change versus the social engineering associated with reducing fossil fuel use in favor of renewable energy, mandating massive human behavioral changes, and redistributing wealth.

> *"If the science is indeed as unassailable as many contend, then why are they so sensitive to criticism and debate?"*

The author of the widely read book entitled, "The Skeptical Environmentalist," Bjorn Lomberg, believes there are better ways to attack global warming than reducing carbon dioxide emissions. His finding: The most cost-effective and technically feasible approach is through geoengineering, the use of technology to deliberately alter the earth's climate.[137] An approach he finds interesting is called marine cloud whitening, wherein fleets of robot ships would pump seawater droplets into the clouds so the clouds will reflect more sunlight back into space. The concept is in its infancy, so its costs and environmental effects are not yet established, but the approach could certainly produce results more rapidly than worldwide CO_2 control.

Another geoengineering technique is carbon sequestration, which is a matter of capturing CO_2 from various sources and burying it underground. As mentioned elsewhere in the book, one place to bury CO_2 is in old oil fields, producing additional oil via Enhanced Oil Recovery (EOR).

Yet another technique involves the launching of large quantities of sulfur aerosols into the atmosphere, which is what volcanoes have done. Indeed, nature provided a large-scale demonstration of this approach in 1991, when Mount Pinatubo erupted and the associated large-scale atmospheric spreading of sulfur aerosols caused noticeable global cooling over the next few years. According to early work by Edward Teller and colleagues, this approach might cost a few billion dollars per year, and the effects would decay in a matter of years, if it was decided to terminate the effort, i.e., it is a technique that could be turned on and off relatively quickly at will. Research on this technique has more recently been

> *"The most cost-effective and technically feasible approach is through geoengineering, the use of technology to deliberately alter the earth's climate."*

pursued by Paul Crutzen, Nobel Prize winner for his work in atmospheric chemistry.

Other techniques have also been suggested. Since a number of these approaches hold promise of relatively low cost and rapid implementation, we find it unsettling that a major research effort on these options has not been adopted and that many environmentalists even reject discussion of geoengineering.

F. The Methane Option

The "Kill Carbon Dioxide" movement suffered a major setback when the 2009 International Copenhagen Climate Change Conference failed to reach a substantial agreement to limit worldwide CO_2 emissions. Subsequently, two climate change luminaries shocked us with an alternate proposal for climate modification. In an editorial in the Wall Street Journal, [138] Robert Watson and Mohamed El-Ashry suggested reducing methane emissions as a means of global warming control. Watson was formerly chair of the IPCC, while El-Ashry is a senior fellow at the United Nations Foundation.

In their editorial, Watson and El-Ashry said the following:

> So what can we do to effectively buffer global warming? The most obvious strategy is to make an all-out effort to reduce emissions of methane. Sometimes called the "other greenhouse gas," methane is responsible for 75% as much warming as carbon dioxide measured over any given 20 years. Unlike carbon dioxide, which remains in the atmosphere for hundreds of years, methane lasts only a decade but packs a powerful punch while it's there.

> This is a huge missed opportunity, and not just for the climate. Methane also forms ozone, the smog that severely damages food crops and kills tens of thousands each year by worsening asthma, emphysema and other respiratory diseases

> Experience has shown that even with modest incentives, methane projects, which are typically small scale, can move fast.

> Scientific studies, such as the EPA's June 2006 report, "Global Mitigation of Non-CO_2 Greenhouse Gases," conservatively indicate that we could eliminate 1.3 gigatons of annual CO_2 equivalent emissions—that's half the U.S. power industry's emissions—just by targeting landfills, coal mines, and oil and gas leaks.

Why was this approach not tabled a long time ago, since it offers a faster, lower cost control option than carbon dioxide emissions limitation?

> **"It appears that the climate change movement was been hiding a potentially attractive alternative to carbon dioxide reduction."**

It appears that the climate change movement has been hiding a potentially attractive alternative to carbon dioxide reduction. Does the withholding of attractive, viable alternatives represent responsible policy behavior?

G. Come on!

In the 1990s when temperatures around the U.S. seemed higher than normal, we were told that we were experiencing global warming. In the winter of 2009-2010, we experienced record-breaking snow and cold weather along the U.S. east coast and in the south. Indeed, we were told that on February 12, 2010, snow existed in every U.S. state other than Hawaii for the first time in recorded history. In addition, as noted, the decade of the 2000s saw essentially flat global temperatures

Some in the climate change community claimed that these contradictory phenomena are all manifestations of global warming. [139] Some even go so far as to contend that any type of weather phenomena is proof of global warming [140] – a claim that sounds more like religion than science. We recognize that there are a number of complicated phenomena and feedbacks in climate science, but it does not seem credible to us that record snow, cold, and a long period of no temperature change can all be evidence of global warming.

H. Global Cooling

In the process of conducting research on climate change, we reviewed a topic that was a concern in the 1970s -- global cooling. Like many others, we were tempted to ignore the subject. However, when we analyzed the technical literature, we found that there were some credentialed technical professionals working on the subject, as evidenced by credible references. [141]

Recognizing that some might reject any discussion of global cooling as out of order, we nevertheless felt it worth noting some of the dangers of global cooling, since the implications of such an event are potentially very worrisome, possibly even more serious than those of global warming. [142]

In the past, colder, harsher weather has led to great famines and major epidemics. The last time the world faced this type of disaster was several centuries ago, and the impacts were devastating – even though they are dim in the collective memory. [143]

Some observations from global cooling events in Europe include:

- The onset of these conditions can be abrupt and severe.
- Food production can decline due to increases in days with overcast skies, a decline in the intensity of sunlight, a decline by several degrees in global temperature, regions of massive rainfall and flooding, some regions of drought, and shortened growing seasons.
- Famines can occur.
- Increased malnutrition can lead to weakened immune systems and influenza epidemics.
- Some glaciers can advance.
- Demand can increase for energy and natural resources.

> *"It does not seem credible to us that record snow, cold, and a long period of no temperature change can all be evidence of global warming."*

Global cooling can delay the start of the planting season in the spring and produce early frosts in the fall, limiting harvests and food production. The cold during the Little Ice Age brought massive crop failures, food riots, famine, and disease, and the Little Ice Age was one of the causal factors of the French Revolution. [144]

Global cooling could disrupt the major grain producing areas of the Northern Hemisphere, including Canada, the northern U.S. Great Plains, northern Europe, Russia, and northern China.[145] Agriculture is already highly petrochemical intensive, and to maintain or expand harvests in the face of global cooling will require more oil and energy – not less.

More generally, the impending decline in world oil production and other energy constraints might be much more serious in an era of global cooling than one of global warming. For example, more energy is required to heat a structure than to cool it, because the temperature difference between outdoor temperatures and human comfort zone temperatures is far greater on the cold side than on the hot side. The falling temperatures associated with a new ice age would not only produce colder winters but also extend the length of the winter season, requiring greater quantities of energy to ensure survivability and livability.

> *"In the past, colder, harsher weather has led to great famines and major epidemics."*

Transportation – by vehicle, train, boat, or airplane – in a colder climate is more difficult and energy intensive than transportation in a warmer climate, as demonstrated today in northern Canada and Alaska for instance.

One of the primary hazards during a cold winter is the loss of electrical power. For example, on January 26, 2009, a deadly ice storm struck the Midwest and caused power and communications to fail, affecting over a million people. In many regions the electrical power was out for several weeks during freezing temperatures. [146] Similarly, in January and February 2008, the heaviest levels of snowfall and freezing rain in 50 years struck China. Heavy snow and sleet paralyzed transport and coal shipments, and many trains were unable to deliver coal to their electrical power plants. At one point, coal reserves were down to emergency levels, and stockpiles were only sufficient for eight days of power generation. [147]

Some recommendations for coping with global cooling are similar to those related to global warming, including increasing the energy efficiency of structures by, for example, increasing insulation in attics, crawl spaces, and exterior walls, plugging air leaks, installing energy efficient windows and doors, and general weatherization procedures.

There is one final set of problems that must be addressed:

- First, the onset of global cooling can be relatively rapid and abrupt. If it has already begun, or is about to, we are talking about effects that may manifest themselves within a decade or two
- Second, the implications of global cooling related to the impending decline in world oil production are ominous, due to the increased energy demands likely to be generated by global cooling
- Third, earlier in this book we noted that mitigation efforts to address declining world oil production must be initiated at least two decades in advance of peak oil to be effective. If global cooling was to begin soon, we may not have two decades to act.

The bottom line is that if the earth is close to the onset of a global cooling cycle, the implications would be very serious. The danger is not a science fiction fantasy of giant glaciers burying New York City and London or the tropics freezing over. Rather, we may be facing relatively minor temperature declines that can have devastating economic, agricultural, and health effects that will be exacerbated by world oil production decline and other energy constraints. It is impossible to assign probabilities to the possible events.

We are not here recommending – as do AGWT advocates – expenditures of hundreds of trillions of dollars on remedial programs. Rather, we here raise the issue and recommend that research and analysis be devoted to it. If global cooling is a potentially serious future threat, then much of the resources and attention over the past few decades devoted to global warming

> **"All these carbon emissions, far from being something frightful, is stopping the onset of a new ice age."**

> **"If the earth is close to the onset of a global cooling cycle, the implications would be very serious."**

may have been wasted.

One final note from Dr. James Lovelock, an independent scientist, author, researcher, environmentalist, and futurist. He is known for proposing the Gaia hypothesis, wherein he postulated that the Earth functions as a kind of super-organism. In March 2010 he shocked colleagues and others when he called for greater caution in climate research. [148] "We're just fiddling around. It is worth thinking that what we are doing in creating all these carbon emissions, far from being something frightful, is stopping the onset of a new ice age." "If we hadn't appeared on the earth, it would be due to go through another ice age and we can look at our part as holding that up." "I hate all this business about feeling guilty about what we're doing. We're not guilty, we never intended to pump CO_2 into the atmosphere, it's just something we did." "They (the skeptics) have kept us from regarding the science of climate change as a religion. It had gone too far that way."

I. Our Conclusions

In light of the above considerations, we have serious questions about climate research and the fixation on global warming as an earth-killing disaster. There are clearly many competent, honest researchers involved in this area of research, but there are also a number that have behaved poorly. Until the discipline undergoes a major restructuring, there is just too much that is too unsettling for us to believe that a scientific basis exists to justify draconian, expensive, civilization-changing measures.

Finally, we believe that the impending decline in world oil production will impose hardships that could be catastrophic to human wellbeing. To effectively mitigate the enormous oil shortage problem while also trying to reduce world carbon dioxide emissions is impossible in our judgment.

> **"To effectively mitigate the enormous oil shortage problem while also trying to reduce world carbon dioxide emissions is impossible in our judgment."**

XVIII. How You Might Be Impacted & What You Might Do

A. The Shock Ahead

In Chapter VIII we described our view of how the shock associated with the decline of world oil production might unfold. In essence, we expect reactions similar to what occurred in the "oil crises" of 1973 and 1979. In many ways, we expect that "It will be the same this time," but it will last much, much longer. Contrast that with what is often heard when discussing difficult situations: "It will be different this time."

B. What options do you have?

> *"It will be the same this time," but it will last much, much longer."*

We anticipate that this book will be published before the sudden, widespread public realization of the decline of world oil production. Our view is that the onset of decline will likely begin in a matter of years. On that basis, you have an opportunity to work through the emotional readjustment that most people feel when they truly comprehend the implications of the world oil production decline problem. It is our experience in talking to various audiences in recent years that most people's first reaction to the oil decline story is one of disbelief and mental shutdown. Thinking about how the world could change due to oil production decline is emotionally and intellectually very difficult. It took each of us time to emerge from the resulting disorientation and to be able to more clearly contemplate what is likely and how we should manage things in our personal lives.

As you get down to serious consideration of what is coming, you will need time to contemplate how you and those close to you can act in an enlightened manner. "To be forewarned is to be forearmed." We hope that you will develop an enlightened, pragmatic plan. On the other hand, you might choose to ignore the problem, as some people do when confronted with overwhelming challenges; that reaction has been called "The deer in the headlight syndrome".

> *"Thinking about how the world could change due to oil production decline is emotionally and intellectually very difficult."*

If you are action-oriented, you have at least two options:

1. Take action soon to get ahead of the problem. You could rearrange your finances, relatively quickly for instance. On the other hand, some options take time, e.g., changing jobs, moving to a more desirable area, purchasing a much more fuel efficient car, etc.

2. Prepare an action plan for the time that widespread public realization begins. Having thought through your options ahead of time, you will be in a better position to take action ahead of the crowd.

C. A List of What Might Be Expected

From the foregoing, here are some of the reactions that we expect will be associated with the sudden, widespread realization of world oil production decline:

- Panic will engulf many people and organizations. There will be a significant emotional shock, disorientation, and feelings of insecurity.

- Immediate liquid fuel shortages are likely as people reflexively top off their gasoline tanks and begin to hoard.

- A dramatic escalation of fuel prices will be associated with large increases in oil prices.

- Difficulty with commutes to work will result from high gasoline prices and growing gasoline shortages.

- Real estate values will decline, maybe plummet in areas located far from work sites or public/mass transportation. Vacation and entertainment areas will be hard hit.

- Considerable decline in stock markets will result from fear, panic, and uncertainty.

- A rapid onset of recession will begin and deepen each year until effective mitigation takes hold.

- Inflation will increase associated with much higher oil prices and their cascading impact on the economic system.

> *"Panic will engulf many people and organizations. There will be a significant emotional shock, disorientation, and feelings of insecurity."*

- Many businesses will stop expansion and likely begin to throttle back, creating growing unemployment.

- Decline in world trade in many goods and services will result from smaller disposable incomes.

- There will be growth in the businesses that are clearly part of the solution – There are always some winners in crises.

This is just a partial list. There will likely be other reactions, because human responses are complex, particularly in disorienting situations.

One of the largest unknowns will be the actions that governments will undertake. In the U.S., we suspect that an administration and congress will likely try to soften the immediate problems, but government options for truly effective action will be limited. On the other hand, unfortunately, there are a number of actions that governments might undertake that would be disastrous, such as oil and oil product price controls, mandating technology winners, etc.

A knowledgeable and pragmatic friend speculated that there will likely be a public hunt for "the guilty." [149] "Since scapegoating often occurs in crisis situations, our friend expects "attacks on the oil traders, oil companies, OPEC, oil distributors, and others. Previously in world history, shortages have led to people being attacked or executed for hoarding. Some politicians will try mightily to shift blame." Our friend is on record as saying that "the worst of America will come out before the best. We will have big inflation. Ownership of oil will be one of the few ways to preserve wealth, but they will tax us on it." We hope our friend is wrong, but we suspect that he is at least partially, if not totally correct.

These are some of the harsh realities that are likely to unfold. We hope for the best, but, as pragmatists, we believe in planning on the grim and the painful-to-contemplate.

D. Some Upsides and Opportunities

Some companies and organizations will flourish in a world of declining oil production. Because physical mitigation will be essential, we expect a major construction boom as companies institute enhanced oil recovery in older oil fields, expand heavy oil development, and build coal-to-liquids and gas-to-liquids plants. Heretofore restricted geographical areas will be opened for oil exploration, and there will be a scramble to develop resources in those areas. Permitting and licensing will happen very quickly, so companies can move to rapid development.

Much more fuel-efficient automobiles will be desirable, but we do not

> **"As pragmatists, we believe in planning on the grim and the painful-to-contemplate."**

expect huge profits associated with their sales, because the oil-decline-induced recession will dramatically reduce the public's ability to purchase new vehicles. This is what happened during the recent "great recession." In the U.S. in 2009, new vehicle sales dropped from a rate of 15-16 million vehicles per year prior to the recession to a rate of less than 10 million vehicles per year.

Research and development on energy efficiency and new energy sources will flourish, largely supported by governments.

Of the renewable energy options, cellulosic biomass may hold the greatest promise of providing useful quantities of liquid fuels with reasonable energy return on energy invested (EROEI) and liquid fuel return on liquid fuel investment (LFROI). While long researched and not yet commercial, we expect enhanced research and development. Algae is a potentially attractive source of liquid fuels, but it has been so for many decades. Technical breakthroughs will likely be required for both cellulosic and algae conversion, but breakthroughs are rarely predictable.

There may be opportunities for companies to make money in deploying technologies and equipment that provide increased energy efficiency. However, in many cases it is likely that people and organizations will continue to operate existing equipment, because of the cost of replacements and their own difficult financial circumstances.

E. Cash for Clunkers

In the aftermath of the initial decline of world oil production, we forecast deepening recession for more than a decade, because even crash program mitigation will take that long to take hold, as we explained.

What about another cash-for-clunkers program? It's possible that governments might want to institute such programs to incentivize people to trade in their low efficiency automobiles for much more efficient ones. However, many governments have already accumulated so much debt that new cash-for-clunkers programs will be very difficult to afford on any kind of meaningful scale. We conclude that the likelihood of big savings in vehicle fuel efficiency will probably be relatively small after the onset of the decline in world oil production, in spite of its obvious attractiveness. Most of the lessening demand for vehicular fuel will come from foregoing personal and business trips, car-pooling, and telecommuting.

F. Some Thoughts on Financial Planning

While we have given a great deal of thought to the world of oil shortages and very high oil prices, we are not experts in financial planning, so the following thoughts should be taken as our personal thinking. In preparation for declining world oil production, each of us acted differently because of our differing circumstances. There is no "one size fits all."

> **There is no "one size fits all."**

Among principles that we believe may be valid are the following:

1. **Avoid holding long-term bonds.** In the high interest rate environment we envision, the value of bonds will decline, possibly significantly. Holding bonds until maturity will recoup face value but the value of the currency will likely be worth much less, so a great deal of value will have been lost due to inflation, particularly for longer-term bonds.

2. **Consider investing in TIPS.** U.S. Treasury Inflation-Protected Securities (TIPS) are government issued bonds that provide a hedge against inflation. The interest rate in TIPS is set when the bonds are sold, but the bond's underlying principal rises and falls with changes in the inflation rate. The intent of these bonds is clear, but they have not yet been tested under severe conditions, so some uncertainty exists.

3. **Consider buying a high mileage automobile early.** Demand for high mileage vehicles was recently relatively modest and the value of existing, lower-mileage cars has been higher than it will be after world oil production goes into decline. It may be a good time to buy a more fuel-efficient car, if you can afford it.

4. **Convert variable interest loans to fixed interest.** As inflation heats up, variable interests will escalate.

5. **Consider delaying paying off long-term, fixed interest loans.** Waiting means that loans will be paid off with lower value currency.

6. **Invest in residential rental properties near local mass transit.** The demand for such properties will likely escalate, since mass transit will allow people opportunities to retain or find employment in far-flung areas.

7. **If your situation permits, move closer to your job and to neighborhoods with better pedestrian access to stores.**

8. **Stay away from very large houses.** The average family dwelling

space in the U.S. and elsewhere will likely continue to trend down and very large homes will likely be more difficult to sell after oil decline.

9. **Consider converting your home heating away from oil or propane**, both of which will increase significantly in cost.

10. **Invest in commodities that are likely to retain their real value in an inflationary environment.** Gold is one option that might be an appropriate investment, since it's been a demonstrated inflation hedge in the past. Certainly there are others, but some are not easy to invest in.

11. **Be wary of investing in stocks of consumer companies.** In a deepening recession, companies that depend on consumer spending are likely to experience lower profitability, as people cut back on their expenditures.

12. **Consider short sales in the stock market.** For those who are comfortable with such transactions, there will likely be a myriad of opportunities. Timing is critical in such transactions.

13. **Consider holding cash or short-term financial instruments.** In the current financial environment, interest rates are very low, but when inflation hits, interest rates are likely to escalate dramatically. Being flexible near-term should allow you to capture higher interest rates later on.

14. **Consider equity holdings in countries that are energy secure.** In principle this is a good concept, but worldwide recession is likely to impact even the most well off, so considerable thought will be required and timing may be very tricky.

These are a few thoughts for you to consider. There are undoubtedly arguments pro and con for each of them, and there are surely other situations to avoid and embrace. The bottom line is that you need to think outside the normal "box" and not be afraid to make significant changes from business-as-usual thinking.

G. Post Script

In the course of making a number of presentations on the future of world oil production, we have had the opportunity to speak to some of the major organizations that provide investment advice to wealthy people and institutional investors. In our experience, few financial advisors have seriously considered the impending decline in world oil production and the ensuing financial environment.

Do not be surprised if you do not find many advisors with clearly thought through scenarios and related financial advice. Of course when the problems become obvious, then everyone will have advice to provide, whether or not they know what they are talking about.

XIX. Conclusions

A. The World Energy Mess As We See It

The title of this book is "The Impending World Energy Mess." "Mess" has a number of meanings that fit the world energy situation: "untidy," "disordered," and "embarrassing confusion." In the foregoing, we described the following messy situations:

- The impending decline of world oil production, which the world has yet to fully comprehend, let alone prepare for.

- The intoxication with renewable energy, which is incapable of providing the world with large amounts of liquid fuels or electric power at costs comparable to alternatives.

- Climate change, the consideration of which has overshadowed a multitude of energy decisions in spite of the fact that there are large questions in the science and that alternative responses have not been given adequate consideration.

The developed world has benefitted from abundant, low cost energy for more than a century, which has resulted in people not appreciating its fundamental importance to our very existence. We are reminded of the Aesop fable:

A man and his wife possessed a goose that laid a golden egg every day. Though fortunate, they decided that they were not getting rich fast enough. Imagining that the bird must be made of gold, they decided to kill it so they could obtain the whole store of precious metal at once. When they cut the goose open, they found its insides to be no different from any other goose, and they lost their treasure.

The world has not lost its golden energy goose, but the goose is in pain. Oil shortage induced human misery may be imminent, and regaining an adequate, balanced energy future will be a long, difficult, and expensive task. There will be no quick fixes because of the inherently large scale of the energy infrastructure.

"Mess" has a number of meanings that fit the world energy situation: "untidy," "disordered," and "embarrassing confusion."

B. World Oil Production

The most serious energy mess facing the world is the impending decline in world oil production. The warning signs include the six-year long plateauing of world oil production, the escalation of oil prices, and the analyses of a number of highly trained professionals and competent organizations. The problem has been called "peak oil" but we prefer to refer to it as the decline in world oil production, because world oil production has been and is likely to stay on the current fluctuating world oil production plateau for a few more years before the onset of production decline. "Peak oil" has been covered to a degree in the media for years, but it has yet to rise high in the public consciousness.

Some in the media describe peak oil and the decline of world oil production as "running out of oil," implying that the next day there will no oil. As we explained, oil is not about to suddenly disappear, rather, world oil production appears ready to go into decline, which means less and less available year after year.

> *"Oil is not about to suddenly disappear, rather, world oil production appears ready to go into decline, which means less and less available year after year."*

How often have you heard of impending disasters with the assertion that "Things will be different this time"? For investors, that claim usually raises a red flag of caution, because often situations are not different. In the case of the impending decline in world oil production, we believe that "Things will be the same this time," except they will last much longer.

The basis for this opinion is our experiences associated with the 1973 Oil Crisis and the Iranian Revolution of 1979. In both cases governments and the public were shocked by the realization of impending oil shortages. It was the shock that caused the public panic that caused people to rush to top off their automobile gasoline tanks and to begin to hoard. Those actions quickly drew down the gasoline supply system and created immediate shortages. If people had not panicked, the supply system would have continued to operate normally for weeks until the last oil tankers unloaded their imported oil cargos. The shocks led to economic distress and public panic along with a

> *"Things will be the same this time, except they will last much longer."*

> **"In the 1973 and 1979 oil crises, there were "valves" to open to quickly alleviate the shortages. In the case of the impending decline in world oil production, there will be no valves to open."**

number of other impacts, which we described.

Keep in mind that in the 1973 and 1979 oil crises, there were "valves" to open to quickly alleviate the shortages. In the case of the impending decline in world oil production, there will be no valves to open. The reason is simple and undeniable; oil is a finite resource – there is only so much in the earth. Thankfully, oil production will not stop suddenly. If oil production stopped abruptly, civilization would go into extreme chaos in short order.

As we described, all oil fields reach maximum production relatively quickly; sometimes they linger on a production plateau; most of the time they do not. At some point all go into a long production decline. History is replete with cases where oil production in countries behaves the same way. As day follows night, world oil production will behave similarly, like it or not.

Most analysts agree that economic damage is inevitable when world oil production declines at a significant rate. Fortunately, the world has a number of administrative and physical mitigation options. However, even if all are implemented quickly and effectively, oil shortages will almost certainly grow more rapidly than our best mitigation efforts, and this implies significant world oil shortages before our mitigation efforts catch up to and surpass world oil production declines. In effect, the problem will run away from our best efforts for a number of years.

Some organizations and individuals are optimistic about future world oil production and believe that there will not be an oil shortage problem for a very long time. The dispute regards the timing, and some imply that business-as-usual will avoid serious problems no matter what. We wish that they were right, but we believe the likelihood is very small.

Perhaps the best way to think of the world oil production decline problem is as a matter of risk. We believe that the risk of widespread economic disaster is so great that immediate action is mandatory. Education, mental rebalancing, and unprecedented action will be required to address the world oil production decline problem.

> *"The risk of widespread economic disaster is so great that immediate action is mandatory."*

C. The Intoxication of Renewable Energy

The concept of renewable energy with its promise of eternal energy from nature is compelling in the abstract. The problem is that this promise is different from many realities for the technologies we have today. More importantly, the realities are very costly and insufficient to satisfy our overall energy needs, let alone our liquid fuel needs.

We discussed wind, solar cells, and biomass-to-liquids. With respect to wind, each of us knows that the wind does not blow all the time, and when it blows, it is at varying speeds. Those facts mean that wind cannot provide us with the reliable power-on-demand that we require in our everyday lives. Practically speaking, the mismatch between wind's natural variability and our electric power needs means that wind energy requires almost 100 percent fossil fuel backup, which more than doubles the cost of wind power to the consumer. Wind energy development is highly subsidized and is mandated by many government jurisdictions. When subsidies were allowed to lapse in the U.S., wind energy construction dropped dramatically, indicating that wind energy is not economic without subsidies.

A similar situation exists with solar cells, which produce electric power at a rate determined by the intensity of light from the sun. Thus, solar cells produce no electric power at night and produce only low levels of electric power when it's cloudy or raining. Solar cells also require 100 percent fossil fuel generation backup. Again, there are government mandates and subsidies that support the installation of solar cell power systems, indicating that they are basically uneconomic.

And then there is the ultimate irony. Taxpayers are providing subsidies to support the deployment of wind and solar technologies, which will inherently raise the price of the electricity that consumers (taxpayers) will consume in the future. The current situation is masochistic, because consumers want reliable electric power at the lowest possible costs, yet they are subsidizing the installation of technologies that will require them to pay more in the future.

We discussed biomass-to-liquids, which could be a potentially attractive way of producing liquid fuels to help alleviate the impending shortages of world oil production. We saw that the land requirements are huge to produce meaningful quantities of biomass-based liquids. Biomass-to-liquid fuels have the inherent problem of needing liquid fuels to grow liquid fuels, so a

> **"Taxpayers are providing subsidies to support the deployment of wind and solar technologies, which will inherently raise the price of the electricity that consumers (taxpayers) will consume in the future."**

large multiplier of liquid fuel produced compared to liquid fuels consumed is essential. In the special case of ethanol from sugarcane in Brazil, the situation is attractive. In the U.S., the corn-to-ethanol mandate is at best marginal with respect to energy balance, and the environmental impacts are not trivial.

Corn-to-ethanol enjoys large subsidies and mandates in the U.S., so taxpayers are paying for the deployment of a marginal, if not losing, energy crop. This is a crop that is also competing for natural resources like land and water against our food crops, which are a basic requirement of human life on the planet.

We hope that current and future research will bring forth attractive renewable electric power and liquid fuel production options. They are just not here today.

The world is intoxicated on renewable energy, but unfortunately, it is more like rotgut than the good stuff we sorely need. Sadly, our infatuation with renewables has caused us to divert our attention from deploying the technologies and fuels that are demonstrated to work effectively and economically.

Education and pragmatic thinking will be required to change what has become an unfortunate fixation.

D. The Global Warming Mess

A number of scientists and politicians are clamoring for the world community to take immediate action to forestall what they describe as the impending devastation of the planet due to global warming. One problem is that the underlying science is extremely complicated, and it is in its relative infancy. Another problem is that political and environmental pressures, and the lure of significant research grants, appear to have skewed the scientific research in favor of dreadful forecasts. As a result of the so-called "Climategate" scandal, we recently discovered that a number of prominent researchers were manipulating their data and results to provide conclusions that fit their biases or their funding sponsors, rather than the science.

> *"The world is intoxicated on renewable energy, but unfortunately, it is more like rotgut than the good stuff we sorely need."*

The world has gone for more than a decade with little change in temperature – no warming, even though the models that predict huge increases in global temperatures by the end of the century did not forecast the recent temperature flattening. Remarkably, those models assumed much larger future oil and coal production than is likely, which means that their forecasts of future temperature increases are much higher than they should be.

Politicians, non-profit organizations, and individuals have hijacked a serious scientific endeavor (climate research) and politicized it for reasons that appear to range from real fear of devastating warming to religious beliefs that humans are moral sinners for creating carbon dioxide. Indeed, the political pressures are so extreme that an individual or organization that questions the science is often branded an undesirable and excluded from what should be honest open scientific discussions.

Finally, when the 2009 Copenhagen climate conference failed to reach a consensus on restricting the emissions of carbon dioxide, some climate luminaries told us that controlling methane emissions could provide a much less expensive and faster solution than carbon dioxide emission reduction. If cheaper and faster is an option, what has kept it from being openly considered in years past?

The science is complicated, and a number of top-notch researchers are involved in developing basic climate science and building models to forecast what might happen in the longer term future. We do not know what the realities of global warming will turn out to be, and we do not expect credible answers until much more research is performed. However, we can tell when we are listening to skewing by special interests and less-than-objective

> *"Politicians, non-profit organizations, and individuals have hijacked a serious scientific endeavor (climate research) and politicized it for reasons that appear to range from real fear of devastating warming to religious beliefs that humans are moral sinners for creating carbon dioxide"*

scientific discourse. It will be some time before the current climate research system is properly readjusted, and credible research brings us to the point of solid understanding.

No matter what the science of climate change tells us, we believe that the decline of world oil production will overshadow climate concerns and will become humankind's most urgent priority, because of the immediate human pain and misery that will ensue.

> *"The decline of world oil production will overshadow climate concerns and will become humankind's most urgent priority."*

E. Some Final Thoughts

We encourage our readers to give these issues careful thought. There is considerable complexity involved, and good answers do not always materialize quickly. We encourage you to consult the huge literature that is available on the Internet under such topics as peak oil, world oil production, energy, conservation, energy efficiency, and the myriad of related topics. Obvious answers and directions often do not emerge quickly, because the issues are complicated and many voices have their own agendas.

With respect to the impending decline in world oil production, we urge you to seriously consider your own personal circumstances and develop a plan for how you will meet the challenges that lie ahead. We also urge you to demand that your governments seriously and explicitly consider these issues and implement intelligent, informed action.

> *"We have faith that humankind will rise to the challenges ahead and will prevail."*

We have faith that humankind will rise to the challenges ahead and will prevail. In the process, the collective "we" will emerge stronger and more pragmatic.

We wish you well.

Bob, Roger, and Bob

Post Script – The Deepwater Horizon Oil Drilling Disaster

As we were completing this book, a terrible oil field accident occurred. The BP Deepwater Horizon drilling rig exploded, costing the lives of eleven workers and unleashing an uncontrolled flow of oil into the Gulf of Mexico. The ensuing environmental damage increased by the day in spite of efforts by BP and the government to stem the flow and minimize the damage. Subsea cameras showed uncontrolled oil flow, and pictures of oil coated birds, turtles, and wetlands filled the media, understandably fueling widespread outrage.

The public demanded that the Obama Administration do something to stop the escaping oil and protect the environment. However, the government does not have the capability to manage complex, high technology oil field operations, but it could and did marshal ships, boats, and personnel to try to contain the environmental damage and begin the clean up.

In an effort to minimize the possibility of another Deepwater Horizon-like occurrence in the future, the Administration announced a number of actions:

1) A commission to review the activities that led to the disaster,
2) A tightening of permitting and regulations for offshore drilling,
3) A six month suspension of drilling in water deeper than 500 ft and the cancellation or temporary suspension of pending lease sales and drilling in the Gulf of Mexico, offshore Virginia, and the Arctic, and
4) A call for cancellation of oil company tax advantages and the acceleration of alternate energy development to move the U.S. off of its dependence on oil.

How these and future actions will play out was not knowable as we completed our book. Nevertheless, it was possible to make some observations about the implications of the aforementioned:

1) The investigative commission will almost certainly identify failures, mishaps, and acts of nature that led to the accident. The information obtained will be useful to both industry and government in their efforts to minimize the reoccurrence of such a tragedy in the future.

2) Even without changes in government regulation, costs for deep offshore drilling will escalate. Insurance rates will increase dramatically, and additional industry-determined safety steps will be adopted, increasing operating costs. More restrictive government permitting and operating regulations will almost certainly increase costs still further.

3) If the suspension of exploratory drilling and other offshore operations is not quickly relaxed, drilling rigs and personnel contracted for offshore oil and gas projects will move to other areas of the world where active drilling opportunities exist. Once this heavy equipment and personnel have moved, it will be difficult and expensive to attract them back to the U.S.

4) It is doubtful that consumers will see any near-term gasoline or diesel price increases, because there will not be a noticeable "jolt" to the oil product supply system, which involves oil flows from a myriad of sources. As the lost production builds over time, increased oil imports will make up for the difference, at least until world oil production goes into decline.

It is our view that there will be little near-term impact on U.S. consumers of oil products. Many in the public may feel satisfied that the government has taken decisive action to minimize the risk of another such offshore accident. Among other things, lawyers will benefit because of the myriad of lawsuits, and the states bordering the GOM will lose a significant number of oil field jobs, if the moratorium stays in place for very long.

While we cannot now know how severe the new government offshore oil and gas restrictions will be, significant actions could well decrease the U.S. capability for offshore oil and gas exploration and production well into the future. Related weakness will exacerbate the U.S. position, when world oil production goes into decline, which we believe will happen within the next five years.

The Energy Information Administration (EIA), the U.S. government's energy forecasting agency, will have to revise downward its long-term forecasts for U.S. oil production in light of the likely reductions in Gulf of Mexico oil production – one of the few bright spots EIA identified for potentially increasing U.S. domestic oil production. Oil imports will increase to make up for the loss, making the U.S. marginally more dependent on imports.

Talk of rescinding tax breaks for oil companies in order to devote more government money to renewable energy typically comes from people who do not understand the messages of this book, specifically that 1) oil will be needed for many more decades; 2) renewable energy cannot provide significant liquid fuel replacements any time soon; and 3) oil shortages cannot be replaced by electric power generation until end-use equipment (automobiles, trucks, buses, trains, etc.) is replaced; which will take decades, as we described.

Lastly, it is worth reflecting on why oil companies are drilling in deepwater to find and produce oil that is so hazardous and expensive. It's because they have no other choice, because there are few good opportunities available to them elsewhere. The places where there is or might be low cost oil are almost completely controlled by national oil companies, which are managing and producing their oil as they see fit. Most of the world's easy-to-find-and-produce oil has been found, a precursor to the onset of world oil production decline.

Footnotes

1) Petrobras's 5-year plan targets large production expansion. OGJ. April 20, 2009
2) These structures are evident in most parts of the U.S. but not in parts of the northeast, where the rocks are mainly granite, which does not show obvious layered geological structures.
3) The word "reserves" is typically used. The word "reserve" is not used.
4) Gbbl stands for giga-barrels, which are billions of barrels.
5) Simmons, M. Twilight in the Desert. Wiley. 2005.
6) For example, Indonesia, an early member of OPEC, became a net oil importer in 2006.
7) Note that oil discovery data are almost always sporadic. For simplicity, discovery data are often smoothed.
8) Mikael Höök, Global Energy Systems, Uppsala University
9) Heads in the Sand. Governments Ignore the Oil Supply Crunch and Threaten the Climate. Global Witness. October 2009
10) NGLs and ethanol have a lower energy content than conventional oil, which is important in some detailed considerations. Also, different liquids are used for non-transportation purposes, such as chemical feedstocks and fuel blending. Refineries produce an array of products from different grades of conventional oil, and these and other hydrocarbon streams flow into different products, depending on local and national considerations. Incredible complications can arise when the disposition of specific liquid fuel streams are considered in detail, but that detail is not important in the broader picture.
11) IEA 2008 World Energy Outlook.
12) Sorrell, S. et al. Oil futures: A comparison of global supply forecasts. Energy Policy. To be published.
13) World Economic Outlook Update. Global Economic Slump Challenges Policies. IMF. January 28, 2009.
14) EIA. Weekly All Countries Spot Price FOB Weighted by Estimated Export Volume (Dollars per Barrel) 10/28/2009.
15) Facing the Hard Truths about Energy - A comprehensive view to 2030 of global oil and natural gas. National Petroleum Council. July 18, 2007
16) The list of contributors to the "peak oil" story is extensive. Some have been pioneers, while others have provided added substance and new insights. Among the notables are the following: M. King Hubbert, Colin Campbell, Jean Laherrere, Matt Simmons, Ken Deffeyes, Jeremy Gilbert, Sadad al-Husseini, Jim Schlesinger, Roger Bentley, Jack Zagar, Jeremy Leggett, Moujahed al-Husseini, Ray Leonard, Chris Skrebowski, Herman Frannsen, David Strahan, and Simon Snowden. Special recognition goes to Professor Kjell Aleklett and colleagues at Uppsala University, including Mikael Höök, Fredrick Robilus, and Bengt Söderbergh. The Association For the Study of Peak Oil (ASPO) has provided useful detailed studies and fostered the open interchange of ideas. ASPO notables include Aleklett, Steve Andrews, Campbell, Rembrandt Koppelaar, Debbie Cook, Dick Lawrence, Jim Baldauf, Randy Udall, and Tom Whipple. Apologies to others not mentioned, who deserve recognition; Lack of mention is not meant as a value judgment. Finally, the three authors of this book have made a number of contributions, hopefully useful.
17) See Robert L. Hirsch, Roger H. Bezdek, and Robert M. Wendling, "Peaking of World Oil Production: Is the Wolf Near?" Geotimes, July 2005.
18) Nashawi, I.S. et al. Forecasting World Crude Oil Production Using Multicyclic Hubbert Model. Energy & Fuels. February 4, 2010.
19) Snedden, J.W. Exploration Play Analysis from a Sequence Stratigraphic Perspective. Search and Discovery Article #40079 (2003)
20) Robelius, F. Giant Oil Fields - The Highway to Oil: Giant Oil Fields and their Importance for Future Oil Production. Uppsala University. 2007.
21) Ibid.
22) Uncertainty about Future Oil Supply Makes It Important to Develop a Strategy for Addressing a Peak and Decline in Oil Production. GAO-07-283. February 2007.
23) Crude Oil – The Supply Outlook. Energy Watch Group. February 2008
24) Heads in the Sand. Global Witness. October 2009.
25) The Oil Crunch-Securing the UKs Energy Future. UK Industry Taskforce on Peak Oil and Energy Security. October 29, 2008
26) The Oil Crunch. A wake-up call for the UK economy. Second report of the UK Industry Taskforce on Peak Oil & Energy Security (ITPOES). February 2010.
27) Sorrell, S. et al. Global Oil Depletion -- An assessment of the evidence for a near-term peak in global oil production. UK ENERGY RESEARCH CENTRE. August 2009.
28) USJFCOM. "Joint Operating Environment 2008." Joint Forces Command, November 2008.
29) Sethuraman, D. DEEP WATER OIL DRILLING SCALED BACK, MAY TIGHTEN CRUDE SUPPLIES. Bloomberg. March 20, 2009.
30) Markey, M. Topic Report: E&P Capital Expenditure Cutbacks. Topic Reports.? Apache Corp. March 16, 2009
31) Cattaneo , C. The Patch: Peak supply vs. peak demand. Financial Post. January 25, 2009.
32) Markey, M. Topic Report: E&P Capital Expenditure Cutbacks. Topic Reports.? Apache Corp. March 16, 2009

33)   SAUDI WARNS OF 'CATASTROPHIC' ENERGY CRUNCH. Reuters. March 19, 2009
34)   Global Energy - 280 projects to change the world. Goldman Sachs. January 15, 2010.
35)   Dr. Schlesinger served three U.S. Presidential Administrations, in addition to being the first Secretary of Energy, he also served as Chairman of the Atomic Energy Commission, Director of Central Intelligence, and as Secretary of Defense.
36)   Höök, M. CERA: Peak Oil is Here." ASPO website. June 10, 2009.
37)   Simmons, M. ASPO's Peak Oil Message: Success and Impediments." ASPO 2009 Interntional Peak Oil Conference. October 12, 2009.
38)   Pickens, T.B. The Plan. PickensPlan.com. July 17, 2008.
39)   Shenk, M. Oil Falls to Four-Week Low on Larger-Than-Expected Supply Gain. Bloomberg. November 12, 2009.
40)   Conoco sees oil output peaking below 100 mln bpd. Reuters. October 20, 2009.
41)   Kelly, R. Chevron CEO Warns On Oil Price; Reiterates Capex Plan. Dow Jones. OCTOBER 18, 2009.
42)   Maxwell, Charles. Energy Outlook. Weeded & Co. September 2009.
43)   Blanco, S. Total CEO's warning: invest now or oil might not be plentiful in 2010. Autobloggreen. September 21, 2009.
44)   Breaking News: Tom Petrie of Bank of America Merrill Lynch says we are at Peak Oil. The Intelligence Daily. September 3, 2009.
45)   Connor, S. Warning: Oil supplies are running out fast. The Independent (UK). August 3, 2009.
46)   Commentary: An Interview with Ray Leonard. Peak Oil Review. ASPO-USA. July 27, 2009.
47)   Ohnsman, A. Toyota to Sell Tiny U.S. 'Urban Commuter' Battery Car by 2012. Bloomberg. January 10, 2009.
48)   Voices of Power: Steven Chu. INTERVIEW OF STEVEN CHU, Secretary of Energy. The Washington Post. April 16, 2009
49)   World Oil Capacity to Peak in 2010 Says Petrobras CEO. The Oil Drum. February 4, 2010.
50)   Whipple, T. Exxon & peak oil. Energy Bulletin. November 9, 2006.
51)   Waldman, P. Exxon vs Obama. Portfolio.com March 18, 2009.
52)   BP: 'We should see volatility increase.' EurActiv interview. www.euractiv.com/en/energy/bp-see-volatility-increase/article-175922. 1 October 2009
53)   Jackson, P. The Future of Global Oil Supply. CERA. November 2009.
54)   Why the "Peak Oil" Theory Falls Down -- Myths, Legends, and the Future of Oil Resources. CERA. November 10, 2006
55)   Lynch, M. "Peak Oil" is a Waste of Energy. New York Times. August 25, 2009.
56)   Lynch, M.C. Peak oil, uncommon ground. ASPO-USA. March 17th, 2008.
57)   Facing the Hard Truths about Energy - A comprehensive view to 2030 of global oil and natural gas. National Petroleum Council. July 18, 2007
58)   Brand, S.R. Herold Pacesetters Energy Conference. ConocoPhillips. September 23, 2008
59)   Sweetman, G. Meeting the World's Demand for Liquid Fuels. EIA. April 7, 2009
60)   Bezdek, R.H., Wendling, R.M. "Fuel Efficiency and the Economy." American Scientist. March-April 2005. pp. 132-139.
61)   Glassman, J.K. "Windfall Profits" Tax on Oil Companies. Capitalism Magazine. September 26, 2005.
62)   The U.S. experience with price and allocation controls during the oil shortages of 1979-1980 is instructive; the historically based allocation of gasoline did not take into account the variation in gasoline consumption reduction across both regions and urban and rural areas.
63)   A 2.3 percent decline in 2010 U.S. GDP would total about $350 billion.
64)   Fiscal drag refers to the deflationary effects of the oil price increases: As consumes pay more for gasoline and other oil products, they have less to spend on other goods and services and the economy suffers.
65)   Oil price drag is similar to fiscal drag and refers to the deflationary effects of higher oil prices.
66)   For example, people may stay closer to home during a shortfall, and the pro rata allocation of gasoline to superhighway stations would oversupply them while undersupplying urban stations.
67)   Other rationing approaches may also be considered. For example, during the Arab oil embargo of late 1973 to mid-1974, license plates ending in either an even or an odd number determined whether you could buy gasoline for your personal vehicle at a service station on an odd or even day of the month. This odd-even tool was also most recently used in Beijing just prior to the 2008 Olympics allowing only those vehicles eligible to be out on the roads. The program successfully improved air quality for the arriving competitors and guests.
68)   See U.S. Department of Transportation, Federal Highway Administration, National Highway Travel Survey, 2004.
69)   Lister, K. et al. Telework Research Network. 2009.
70)   See National Science Foundation, "NSF Press Release 08-038 "Telework Benefits Employers, Employees and the Environment" and "Telework Under the Microscope – A Report on the National Science Foundation's Telework Program", March 11, 2008
71)   Hirsch, R.L. et al. Peaking of World Oil Production: Impacts, Mitigation, & Risk Management. DOE NETL. February 2005.

72) Platts. IEA says demands on OPEC crude to increase 'substantially.' November 7, 2006.
73) EIA. Energy in Brief. January 28, 2009.
74) Hirsch, R.L. Mitigation of Maximum World Oil Production: Shortage Scenarios. Energy Policy. February 2008.
75) Oil Shockwave - Oil Crisis Executive Simulation. SAFE. April 2008.
76) See our discussion of Liquid Fuel Return on Investment (LFROI) elsewhere in the book.
77) http://www.hybridcars.com/hybrid-sales-dashboard/december-2008-dashboard-focus-production-numbers-25416.html#states-total-sales
78) PLUG-IN HYBRID VEHICLE COSTS LIKELY TO REMAIN HIGH, BENEFITS MODEST FOR DECADES. Office of News and Public Information. The National Academies December 14, 2009.
79) Real Prospects for Energy Efficiency in the United States. America's Energy Future Panel on Energy Efficiency Technologies. National Academies Energy. 2009.
80) Soderbergh, B., et al. A Crash Program Scenario for the Canadian Oil Sands Industry. Energy Policy. March 2007.
81) One of the major lessons of the 2007-09 recession is that the first thing to become scarce during a recession is financing to purchase almost anything, much less large, expensive capital equipment with long payback periods.
82) The primary 2008 data have been collected from the U.S. Department of Commerce, Bureau of Economic Analysis; the U.S. Department of Energy, Energy Information Administration and Oak Ridge National Laboratory; and the U.S. Department of Transportation, Bureau of Transportation Statistics.
83) Source: U.S. Department of Commerce, Bureau of Economic Analysis; the U.S. Department of Energy, Energy Information Administration and Oak Ridge National Laboratory; and the U.S. Department of Transportation, Bureau of Transportation Statistics.
84) Technologies and Approaches to Reducing the Fuel Consumption of Medium and Heavy-Duty Vehicles. National Academies Press. 2010.
85) Transitions to Alternative Transportation Technologies – Plug-In Hybrid Electric Vehicles. National Academies Press. 2010.
86) Hall, A.S. et al. Revisiting the Limits to Growth After Peak Oil. American Scientist. May-June, 2009.
87) Ibid.
88) Hall, C.A.S., Cleveland, C.J. EROI: Definition, history and future implications. ASPO-USA Conference. Denver. November 1, 2005.
89) Ibid
90) Economic input-output analysis is often used to develop such estimates.
91) These basics were understood back in the mid 1970s, when one of us (RLH) had responsibility for the federal biomass program in ERDA, the predecessor to the Department of Energy. This concept remains valid and is widely understood.
92) Annual Energy Outlook. EIA. 2009.
93) Aubrey McClendon, CEO of Chesapeake Energy – one of the more aggressive U.S. natural gas companies, contends that shale gas at $5/MCF is not sustainable. The required "sustainable" price may actually be much higher than that. Benjamin Dell, of Bernstein Research in New York, who has conducted some of the most rigorous research on the shale gas industry, estimates that the full cost of finding, developing, and operating shale gas wells, and paying an average return on capital to investors, requires a spot gas price of $7.50 to $8/MCF. Note that these are the prices to producers; the prices to consumers are much higher. See the discussion in John Dizard, "The True Cost of Shale Gas Production," Financial Times, March 7, 2010.
94) Basil Katz, New York. "Ruling Sets Back Gas Drilling in Watersheds." Reuters. April 23, 2010.
95) Robert W. Howarth, "Preliminary Assessment of the Greenhouse Gas Emissions From Natural Gas Obtained by Hydraulic Fracturing," Department of Ecology and Evolutionary Biology, Cornell University, March 2010.
96) Realizing the Energy Potential of Methane Hydrate for the United States. National Research Council. February 2010.
97) Transitions to Alternative Transportation Technologies--A Focus on Hydrogen. The National Academies. 2008.
98) Jaramillo, P. et al. Comparative Life Cycle Carbon Emissions of LNG Versus Coal and Gas for Electricity Generation. Carnegie Mellon University. 2007.
99) Howarth, R. W. Preliminary Assessment of the Greenhouse Gas Emissions from Natural Gas obtained by Hydraulic Fracturing. Cornell University. March 17, 2010.
100) Sperling, D. Producing Liquid Fuels from Biomass. Workshop on Trends in Oil Supply and Demand and Potential for Peaking of Conventional Oil Production. National Research Council Workshop. October 21, 2005. Hirsch, R.L. Knowledge gained while managing the federal biomass program. ERDA. 1976-1977.
101) Oil Shale Fact Sheet. U.S. Department of Energy, Office of Naval Petroleum and Oil Shale Reserves. 2006.
102) Transitions to Alternative Transportation Technologies – A Focus on Hydrogen. The National Research Council of the National Academies. 2008.

246

103) DOE EIA. Recoverable Coal Reserves at Producing Mines, Estimated Recoverable Reserves, and Demonstrated Reserve Base by Mining Method. 2004, Table 15.
104) Ibid.
105) "EIA. Coal Reserves Data. [http://www.eia.doe.gov/cneaf/coal/reserves/chapter1.html#fig2], 1997. The estimates were actually compiled by the USGS.
106) Ibid.
107) Matlack, C. A High-End Bet on Nuclear Power. BuinessWeek. March 15, 2010.
108) International Energy Outlook 2009. U.S. EIA. May 27, 2009.
109) The Future of Nuclear Power. Massachusetts Institute of Technology. 2003; The Economic Future of Nuclear Power. University of Chicago, August 2004; Nuclear Power: Outlook for New U.S. Reactors. U.S. Congressional Research Service. March 2007; Joskow, P.L. The Future of Nuclear Power in the United States: Economic and Regulatory Challenges. AEI-Brookings Joint Center for Regulatory Studies. December 2006; The Economics of U.S. Climate Policy: Impact on the Electric Industry. The Brattle Group. March 2007; Nuclear Power "Renaissance" Moving Beyond Talk to Real Action. Cambridge Energy Research Associates. April 2007. To ensure comparability of the studies' assumption and findings, all costs have been converted to constant 2008 dollars.
110) New Nuclear Plants. Nuclear Energy Institute. February 2010.
111) Dixon, D. Wind Generation Performance During the July 2006 California Heat Storm. EnergyPulse. September 8, 2006.
112) Bryce, R. Texas Wind Power: The Numbers Versus the Hype. Energy Tribune. August 5, 2009.
113) www.transmission.bpa.gov/business/operations/wind/WindGen_VeryLow_Jan08Jan09x.xls
114) Bezdek, R.H. Potential Unforeseen Consequences of Dedicated Renewable Energy Transmission: Implications For the European Energy Market. Presented at the World Energy Congress, Montreal, Canada. September 2009.
115) American Wind Energy Association. 2008.
116) Solar Hero. Solar Today. September/October 2009, p.14.
117) Redell, C. NW Utilities Get Wind of Integration Charge. Reuters. August 12, 2009.
118) ITER. < http://en.wikipedia.org/wiki/ITER>
119) NIF. < https://lasers.llnl.gov/>
120) Inertial electrostatic confinement. <http://en.wikipedia.org/wiki/Inertial_electrostatic_confinement>
121) Our data came from the BP Statistical Review of World Energy for consumption and production and the U.S. Energy Information Administration (EIA) International Energy Statistics database for imports and exports. Other potential data sources are the U.S. Central Intelligence Agency (CIA) and the International Energy Agency (IEA).
122) Unfortunately, wise investment of oil wealth has often been the exception rather than the norm due to what is known as the "resource curse." See the discussion in Maass, P. Crude World: The Violent Twilight of Oil. Alfred A. Knopf 2009.
123) Petre, J. Climategate U-turn as scientist at centre of row admits: There has been no global warming since 1995. Dailymail.co.uk. February 14, 2010.
124) Rogers, M. A Skeptical Take on Global Warming. Washington Post. September 10, 2009.
125) For example, consider the following. "Those odd climatological phenomena led me to reflect on the rapidly changing weather patterns that are altering the way we live. In Virginia, the weather also has changed dramatically. Recently arrived residents in the northern suburbs, accustomed to today's anemic winters, might find it astonishing to learn that there were once ski runs on Ballantrae Hill in McLean, with a rope tow and local ski club. Snow is so scarce today that most Virginia children probably don't own a sled. But neighbors came to our home at Hickory Hill nearly every winter weekend to ride saucers and Flexible Flyers." Kennedy Jr., R.F., Palin's Big Oil Infatuation. Los Angeles Times. September 24, 2008. Note: In the winter of 2009-10 Mclean, Virginia, along with the rest of the Washington, D.C. metropolitan area experienced the most snow in recorded history. "We had daffodils growing in our front yard in Washington this December. It was like being in the Twilight Zone. I half expected Rod Serling to be out mowing my lawn. Now, the bad news: If we don't fix this thing in the next 50 years, we're toast." Thomas Friedman, speech at "Eweek at Stanford," Stanford University, March 2, 2007.
126) Pielke, R. et. al. Normalized Hurricane Damage in the United States: 1900-2005. Natural Hazards Review. February 2008.
127) Rogers, op. cit.
128) Avery, D. 31000 scientists sign Oregon GW Skeptic Petition. Canada Free Press. May 24, 2008.
129) The "climategate" scandal has been scrutinized and documented by numerous sources; for example, see Climategate Document Database, www.climate-gate.org.
130) Booker, C. Climate Change: This is the Worst Scientific Scandal of Our Generation. Telegraph.co.uk. November 28, 2009.
131) http://biggovernment.com/jlott/2010/02/11/the-real-climategate-scandal/
132) Moran, R. Global warming science implodes overseas: American media silent. American Thinker. January 31, 2010.
133) Petre, J. Climategate U-turn as scientist at centre of row admits: There has been no global warming since 1995. Dailymail.co.uk. February 14, 2010.

134)   Krugman, P. Betraying the Planet. New York Times. June 28, 2009.
135)   Pilkington, E. Put Oil Firm Chiefs On Trial, Says Leading Climate Change Scientist. The Guardian. June 23, 2008.
136)   Policy Implications of Greenhouse Warming: Mitigation, Adaptation, and the Science Base (1992), Committee on Science, Engineering, and Public Policy (COSEPUP)
137)   Broder, J. A Skeptic Finds Faith in Geoengineering. New York Times. Green Inc. September 3, 2009.
138)   Watson, R., El-Ashry, M. A Fast, Cheap Way to Cool the Planet. WSJ. December 29, 2009.
139)   For example, McKibben, B. Washington's Snowstorms, Brought to you by Global Warming. Washington Post. February 14, 2010.
140)   Rose, J. Unusual amounts of snow or lack of snow are all signs of global warming in "Ways Around the Impasse." New York Times. February 15, 2010. Mr. Rose is a board member of the Natural Resources Defense Council.
141)   Global Cooling References
N.S. Keenlyside, M. Latif, J. Jungclaus, L. Kornblueh, and E. Roeckner, "Advancing Decadal-Scale Climate Prediction in the North Atlantic Sector," Nature, vol. 453, (May 1, 2008), pp. 84-88.
Qing Bin Lu. "Cosmic-Ray-Driven Electron-Induced Reactions of Halogenated Molecules Adsorbed on Ice Surfaces: Implications For Atmospheric Ozone Depletion and Global Climate Change," Physics Reports, Vol. 487, No. 5, Feb 2010, pp. 141-167
Engvild. Kjeld C. "A Review of the Risks of Sudden Global Cooling and its Effects on Agriculture." Agricultural and Forest Meteorology, vol. 115, no. 3-4, 2003, pp. 127-137.
David Archibald, "Solar Cycle 24: Implications for the United States," presented at the International Conference on Climate Change, March, 2008.
Kent, D.V. and G. Muttoni. "Equatorial Convergence of India and Early Cenozoic Climate Trends," Proceedings National Academy of Sciences, doi:10.1073/pnas. 0805382105. 2008.
Hakan Grudd. "Torneträsk Tree-Ring Width and Density AD 500–2004: A Test of Climatic Sensitivity and a New 1500-Year Reconstruction of North Fennoscandian Summers, Climate Dynamics, Volume 31, Numbers 7-8, December, 2008, pages 843-857.
C. de Jager and S. Duhau, "Forecasting the Parameters of Sunspot Cycle 24 and Beyond," Journal of Atmospheric and Solar-Terrestrial Physics, Vol. 71, No. 2 (February 2009), pp. 239-245.
James M. Taylor. "Global Cooling Continues," Environment and Climate News, March 2008.
M.A. Clilverd, E. Clarke, T. Ulich, H. Rishbeth, and M.J. Jarvis, "Predicting Solar Cycle 24 and Beyond," Space Weather, Vol. 4, 2006.
Don J. Easterbrook. "Global Cooling is Here: Evidence for Predicting Global Cooling for the Next Three Decades," Global Research, November 2, 2008, Department of Geology, Western Washington University.
I.G. Usoskin, S.K. Solanki, and G.A. Kovaltsov, "Grand Minima and Maxima of Solar Activity: New Observational Constraints, Astronomy and Astrophysics, Vol. 471, 2007, pp. 301-309,
Fred Pearce. "World Will Cool for the Next Decade," New Scientist, September 9, 2009.
N. S. Keenlyside, M. Latif, J. Jungclaus, L. Kornblueh, and E. Roeckner. "Advancing Decadal-Scale Climate Prediction in the North Atlantic Sector, Nature, vol. 453, 84-88 (1 May 2008).
David Archibald, "Solar Cycles 24 and 25 and Predicted Climate Response," Energy and Environment, Vol. 17, No. 1 (2006)
Eelco J. Rohling and Heiko Pälike. "Centennial-scale Climate Cooling With a Sudden Cold Event Around 8,200 Years Ago," Nature, Volume 434, Issue: 7036, 2005, Pages 75-979.
How Well do Scientists Understand How Changes in Earth's Orbit Affect Long-Term Natural Climate Trends? Science Daily, Feb. 7, 2010.
Kh. I. Abdusamatov, "Optimal Prediction of the Peak of the Next 11-Year Activity Cycle and of the Peaks of Several Succeeding Cycles on the Basis of Long-Term Variations in the Solar Radius or Solar Constant," Kinematics and Physics of Celestial Bodies, Vol. 23, No. 3 (June 2007), pp. 97-100.
142)   Marusek, J.A. Solar 'Grand Minima' Threat Analysis. June 3, 2009, and Marusek, J.A. "Grand Minima Preparedness Plan. May 2009.
143)   Attempts by the IPCC in the infamous hockey stick graph to eliminate the Little Ice Age, which occurred from the 16th to the 19th centuries, do not negate its existence. See Fagan, B. The Little Ice Age: How Climate Made History, 1300-1850. Basic Books. 2002.
144)   The historian John D. Post referred to this period as the "last great subsistence crisis in the Western world." See Post, J.D. The Last Great Subsistence Crisis in the Western World. Nomilk Books. 1977
145)   Booker, C. Crops Under Stress as Temperatures Fall. The Telegraph. November 19, 2009,
146)   Schreiner, B.and Taylor,B. "Kentucky Ice Storm: Nearly One Million Still Without Power. Associated Press. January 31, 2009.
147)   Jian, Y. and Lydia Chen, L. Crisis in Chenzhou – No Power or Water. The Shanghai Daily. January 31, 2008.
148)   Bowater, D. How Carbon Taxes Have Saved Us From a New Ice Age. Express.co.uk. March 11, 2010, and Clover, C. Granddaddy of green, James Lovelock, warms to eco-skeptics. The Sunday Times. March 14, 2010.
149)   Kummer, L. Private communication.

248